Infection Control in Healthcare

Infection Control in Healthcare

Peter Meers

Formerly Associate Professor
Microbiology Department
National University of Singapore
One time Director
Division of Hospital Infection
Central Public Health Laboratory
Colindale, London, UK

Madeleine McPherson

Infection Control Nurse
Fremantle Hospital
Western Australia

Judith Sedgwick

National Infection Surveillance
Central Public Health Laboratory
Unit Colindale, London, UK

Stanley Thornes (Publishers) Ltd

First published 1997 by:
Stanley Thornes (Publishers) Ltd
Ellenborough House
Wellington Street
CHELTENHAM
GL50 1YW
United Kingdom

97 98 99 00 01 / 10 9 8 7 6 5 4 3 2 1

A catalogue record for this book is available from the British Library

ISBN 0-7487-3318-3

Typeset by Acorn Bookwork, Salisbury, Wilts.
Printed and bound in Great Britain by Scotprint, Musselburgh

Contents

Preface

Hospital Infection Control for Nurses began life in the late 1980s as a set of notes used in tutorials for a newly appointed infection control nurse. This book is its direct successor. The original text has been revised to accommodate changes in the way healthcare is delivered and in the infectious problems that complicate it. Additions have been made in response to obvious need and details have been altered in the light of helpful criticisms received from colleagues. Despite major changes, however, the general approach does not differ from that set out in the earlier preface.

The word 'iatrogenic' is given new prominence to mark the growth of the delivery of healthcare in the community. Changes in practice, notably in surgery, are reviewed. Current and developing difficulties with methicillin-resistant strains of *Staphylococcus aureus*, vancomycin-resistant enterococci, *Clostridium difficile, Mycobacterium tuberculosis* and the prions are noted. The contribution made by antimicrobial drugs to the emergence of some of these and of other iatrogenic pathogens is described. The varieties of infections that are encouraged by the more extreme forms of immunodeficiency, the range of microbes involved, and the risks these pose to healthcare workers are discussed. Attention is paid to the growing importance of clinical audit.

Peter Meers
Madeleine McPherson
Judith Sedgwick
1997

Preface to *Hospital Infection Control for Nurses*

'How to do it' books differ. Some give precise instructions, others describe the tools, and readers are left to use them according to circumstances. Conditions in hospitals vary enormously, so we have followed the latter pattern. Basic information on infection control is provided to allow rational, cost-effective choices to be made, tailored to local need and resources. We hope it will be useful to health professionals working in countries at all levels of socioeconomic development and in any type of healthcare system.

To meet this specification we have examined the roots of current practice. If we find these wanting, we say so. This is not simple anarchy. The average individual can no longer afford lifetime medical care to modern standards. Attempts to make this a collective responsibility through insurance or taxation have not succeeded. The result is rationing. In richer countries this operates either by putting the full range of medical care out of the reach of less affluent citizens or by making them wait months or years for treatment. Poorer countries see the cost of services they would like to introduce rise faster than their national incomes. In these circumstances anything that can reduce the cost of medical care without affecting its quality must be welcome. Of course, we have not found a solution to a problem that has eluded others. We claim only to add savings that may be derived from a critical appraisal of current practices to the significant benefits already available from the application of effective infection control policies.

In assembling our material we have tried to be brief yet comprehensive. There are numerous textbooks on microbiology and infectious diseases. We include here only as much of these subjects as is particularly relevant or necessary for immediate comprehension. We have delved a little into history. We believe this is important, particularly in cases where current practice reflects past errors or misconceptions.

The book is intended for medical and nursing practitioners in infection control and nurse educators. We think much of it might be read with benefit by hospital administrators, medical students, those concerned with the central supply of diagnostic and therapeutic materials and indeed anyone whose duties involve direct or indirect contact with patients.

Peter Meers
Wendy Jacobsen
Madeleine McPherson
1992

Glossary and abbreviations

Words and phrases that have more general meanings are defined here in the context of infection or of medical microbiology. Additional definitions may be traced through the index.

Aerobe A microbe that grows in the presence of oxygen, and not in its absence (see *Facultative anaerobe*).

AIDS Acquired immune deficiency syndrome, due to infection with the human immunodeficiency virus (HIV).

Anaerobe A microbe that cannot grow in the presence of oxygen.

Antibody A protein molecule produced by the body as a result of immunization with an antigen, that unites with its antigen in a highly specific way.

Antigen Any substance which, when introduced into the body, produces an immune response.

Antimicrobial drug A drug used to treat or prevent an infection (**antimicrobial therapy**, the use of such a drug).

APIC Association for Practitioners in Infection Control.

Bacillus A rod-shaped bacterial cell: also the name of a bacterial genus.

Bacteraemia The symptomless presence of bacteria in the blood.

Bacteriology The scientific study of bacteria.

CAI Community-acquired infection.

CDC Centers for Disease Control, NE Atlanta, Georgia, USA (now the Centers for Disease Control and Prevention).

CDSC Communicable Disease Surveillance Centre, Colindale, London, UK.

CIO Control of infection officer.

Colonization A long-term relationship in which a microbe lives on (or in) a host, without any reaction by the host to its presence.

CPHL Central Public Health Laboratory, Colindale, London, UK.

CQI Continuous quality improvement.

CSSD Central sterile supply department.

Cellular immunity Functions of the immune system that depend on the direct action of specialized cells on foreign materials (antigens) in the body.

Colony A collection of microbes of the same type, in one place: as grown artificially in a laboratory to form a visible clump of millions of them.

Commensal A microbe that establishes a colonization without harm to its host.

CNS Central nervous system.

CSF Cerebrospinal fluid.

Culture A process by which microbes are induced to multiply artificially (**culture medium**, a nutrient mixture in or on which culture is performed in a laboratory).

CVS Cardiovascular system.

Deoxyribonucleic acid One form of nucleic acid with a unique chemical structure, found in all living cells; the repository of genetic information.

DNA Deoxyribonucleic acid.

EBV Epstein-Barr virus.

EM Electron microscope.

Enzyme A protein that acts as a biological catalyst in the process of metabolism to effect the breakdown or the building up of the molecular components of cells.

EO Ethylene oxide.

Epidemic See *Outbreak*.

Epidemiology The scientific study of the spread and the control of diseases.

Facultative anaerobe A microbe that can grow either in the presence or in the absence of oxygen.

Flora The assembly of plants or microbes that are peculiar to a particular region or to a more restricted environment, for example, the surface of the human body or a part of it.

Gene A short stretch of DNA, part of a much longer molecule. A gene is the blueprint that determines the precise structure of a molecule in a cell and it is also responsible for the inheritance of this function.

Genetics The scientific study of genes and heredity.

Genus A collection of species of living things with sufficient overall similarity in the detail of their structure for them to be classified together. The **generic name** of a microbe is the word that appears first in its Latin description (see *Species*).

GI Gastrointestinal infection.

GNR Gram-negative, rod-shaped bacteria.

HAI Hospital-acquired (nosocomial) infection.

HBV The hepatitis B virus.

HCV The hepatitis C virus.

HCW Healthcare worker.

HEPA High-efficiency particulate air (filter or filtration).

HIV Human immunodeficiency virus (see *AIDS*).

Host The larger partner in a host–parasite relationship or colonization: in medical usage, a human being.

HSDU Hospital sterilization and disinfection unit.

HSV Herpes simplex virus.

Humoral immunity Functions of the immune system that depend on substances dissolved in body fluids, particularly those concerned with antibodies.

IaI, iatrogenic infection An infection that is an unwanted consequence of a medical intervention, also an occupational infection in a healthcare worker.

ICC Infection control committee.

ICD Infection control doctor.

ICI Intravascular cannula-associated infection.

ICN Infection control nurse.

ICNA Infection Control Nurses' Association.

ICO Infection control officer.

ICT Infection control team.

ICU Intensive care unit.

Immune (Immunity) The state of an individual who has been immunized or is otherwise insusceptible to an infection.

Immune response The reaction of the body's immune system when it is stimulated by some foreign material (an antigen).

Immune system Specialized tissues and their chemical products that operate throughout the body to produce various immune responses.

Immunize (Immunization) The process by which an immune response is generated to an antigen that has been introduced, naturally or artificially, into the body.

Immunocompetent The state of an individual whose immune system is fully operative.

Immunodeficient The state of an individual whose immune system is, to a variable extent, defective.

Immunodepression A reduction in the efficiency of the immune system, due to any cause.

Immunoglobulin Antibody.

Immunology The scientific study of immunity.

Immunosuppression The production of immunodeficiency by artificial means (drugs, radiation, etc.).

Incubation The exposure of a microbial culture to a temperature and gaseous environment designed to encourage growth and multiplication of the microbes concerned (**incubator**, the apparatus that provides this environment in a laboratory).

Infection The outcome of an interaction between a host and a microbe in which the host reacts in an observable way. The result is a **clinical infection** when the individual becomes ill or a **subclinical infection** if the evidence for it can only be detected in a laboratory.

LRI Lower respiratory tract infection.

MIC Minimum inhibitory concentration.

Microbiology The science of the study of microbes.

MID Minimum infectious dose (of microbes).

MARSA Methicillin- and aminoglycoside-resistant *Staphylococcus aureus*.

Microbe A living organism (or a virus), too small to see with the naked eye.

MRC Medical Research Council (UK).

MRSA Methicillin-resistant *Staphylococcus aureus*.

Mycology The scientific study of fungi.

NUH National University Hospital, Kent Ridge, Singapore.

Normal flora The collection of microbes usually present as colonists in any specified environment, including some surfaces of the healthy human body.

Nosocomial infection Synonymous with HAI.

Nucleic acid A long string of individual nucleotide sugars (see *DNA* and *RNA*).

OD Operating department.

OR Operating room.

ORSA Ordinarily resistant *Staphylococcus aureus*, i.e. not MRSA or MARSA.

Outbreak The epidemiological condition that exists when two or more patients suffer from infections that are related to each other as to their cause.

Pathogen A microbe or larger parasite that causes harm to its host and so produces infection or disease (**pathogenicity**, the expression of this property).

PML Polymorphonuclear leucocyte.

Privileged surface A surface of the body equipped with mechanisms that keep it free of microbes (e.g. healthy respiratory or urinary tracts).

Ribonucleic acid A form of nucleic acid with a particular chemical structure. RNA is involved in protein synthesis though in some viruses it has replaced DNA as the repository of genetic information.

RNA Ribonucleic acid.

RTI Respiratory tract infection.

SENIC Study on the efficacy of nosocomial infection control

Septicaemia The presence of microbes in the blood, accompanied by symptoms.

Species A basic division within the classification of living organisms: members of the same species have a very high degree of structural, chemical and genetic similarity. The second word in the full Latin title of a microbe is its **specific name** (see *Genus*).

Spore A resting form produced by certain microbes that is unusually resistant to desiccation and chemical attack: an adaptation to survive adverse conditions.

SWI Surgical wound infection.

TPN Total parenteral nutrition.

TQM Total quality management.

TSSU Theatre sterile supply unit.

UTI Urinary tract infection.

uV Ultraviolet (light).

Vaccination The process of immunization with a vaccine.

Vaccine A preparation of antigen(s) used to produce immunity. Antigens in vaccines are usually derived from or consist of microbes.

Virology The scientific study of viruses.

VRE Vancomycin-resistant enterococci.

VZV The varicella-zoster virus.

About this book

Why the word iatrogenic has been adopted in this book and how its contents are arranged.

Why 'iatrogenic'?

As noted in the preface, this book is a direct descendent of another called *Hospital Infection Control for Nurses*, published in 1992. In the interval practice in healthcare has changed to the extent that the authors had to reconsider the use of the terms 'hospital', or 'hospital-acquired', as applied to the types of infections that were its subject. The conclusion reached was that these terms no longer represent the meaning required and that their inadequacy will increase in the future. The word 'iatrogenic' (derived from the ancient Greek for a healer) is an alternative which, though barely mentioned in the parent volume, now more appropriately expresses our meaning. Although not yet in general use, it has been adopted to avoid the awkward circumlocutions that would have been necessary to extend the meaning of the original terms.

Iatrogenic infections (abbreviated to **IaIs** to avoid confusion with a Roman numeral) are infections that arise as unwanted consequences of interventions applied in the context of healthcare (or more concisely, infections that arise as consequences of medical interventions). Medical interventions may be for diagnostic or therapeutic purposes and range from the simple taking of a pulse through more invasive procedures such as endoscopy, intravenous therapy or radiological contrast studies, to truly major activities such as surgery with extracorporeal circulation.

The new terminology is necessary because the delivery of healthcare is passing through what, even by its own rapidly changing standards, is an unusual state of flux. The seemingly endless and exponential rise in the cost of medical diagnosis and treatment has triggered an urgent and now increasingly political search for ways to stem the tide. Residential treatment is generally the more expensive part of healthcare, so one of the methods applied has been to reduce the time the average patient spends in hospital. As a result the number of hospital beds has fallen and the money saved has been diverted elsewhere. The rising curve of expenditure has flattened out a little, but this is only temporary. There has been no decrease in the number of patients treated, so the effect has been to place a considerable extra burden on the medical services outside hospitals.

Despite the transfer of activity, many iatrogenic infections still originate in hospitals and these may still be called hospital acquired (or nosocomial). It cannot be doubted that the pressure to save money and the proportion of minor medical interventions carried out in the community will continue to rise. It is not yet possible to predict how far this will go, but the number of iatrogenic infections that originate outside hospitals will grow and these are certainly not hospital acquired (or nosocomial). Other implications of the change in terminology from 'hospital acquired' to 'iatrogenic' are discussed in Chapter 3, p.26.

The incubation period of an iatrogenic infection (IaI) is the time that elapses between the event that initiates the process and the appearance of an infection as a clinically recognizable entity. This period is nearly always measured in a small number of days. Not very long ago patients were admitted to hospitals for periods of weeks so there was time for nearly all the infections they acquired as a result of their admissions to become clinically evident before they were discharged. Nowadays the length of the average admission is more likely to be counted in days (and sometimes in hours) rather than in weeks. The result is that a large and increasing proportion of infections, even though they are hospital acquired, now first appear and are dealt with in the community.

Infection control practitioners and community healthcare workers must come to terms with these substantial and still growing changes in the pattern of their daily work. One consequence is that many infections that result from the actions of hospital-based healthcare workers now declare themselves after the patients concerned have moved beyond the range of their clinical experience. It is important to devise a system that keeps those concerned informed of the infections they cause. Because many of these infections now emerge beyond the range of the vision of those who might prevent them, it is even more important to ensure that they are not ignored (Box 1.1).

Box 1.1   Avoidable infections

A computer programmed to translate English to and from a foreign language was challenged to interpret the axiom 'out of sight, out of mind', first in one direction, then back again. After the double transfer the words re-emerged as 'blind idiot'! As a general rule, healthcare workers tend to play down or ignore the accidental harm that sometimes complicates their daily activities, even though, with a little thought, at least some of it is avoidable. There are none so blind as those who (even subconsciously) do not wish to see!

The arrangement of this book

An infection is the outcome of a dynamic process that involves the interplay of a number of separate, individually variable factors. Average citizens (including some who are supposed to know better) do not comprehend or even dream of the complexities involved. They interpret their experiences of infections in the light of oversimplifications, folklore and popular prejudice. This is no basis for the study or effective practice of anything, let alone the control of infection. Science has begun to illuminate the invisible world inhabited by microbes and, in the environment we share, to make sense of the way we interact with them.

Infection control still has about it traces of the incantations and rituals that mark its non-scientific origins. These are an expensive waste of time and they have no place in the world of the 21st century. Science is based on measurement. Effective infection control can only be based on a precise knowledge of how many infections afflict the population under consideration, their varieties, where within the population they are found and a good estimate of the success or failure of procedures introduced to influence them. Infection must be measured if it is to be controlled.

One of the first rules of warfare is to understand your enemy. Infection control is a form of warfare and the background information required for an understanding of IaIs is provided in Chapters 2, 3 and 4 of this book. Chapter 2 sets the scene with a historical introduction that traces the origin of some popular misconceptions and misinterpretations, in the hope that they may be avoided in the future. Chapter 3 discusses the variables involved in the causation and distribution of IaIs and Chapter 4 describes how IaIs are measured (surveyed or audited).

If they are to be cost-effective, procedures for the control of IaIs must be designed in the light of the information provided in Chapters 2, 3 and 4. Procedures applied without comprehension rapidly degenerate into meaningless rituals, increasingly used in inappropriate circumstances. Chapters 5–11 are concerned with the procedures that provide the foundation for the control of IaIs, based on the principles set out in the earlier chapters. Chapter 12 looks to the future of the subject and the Appendix provides a set of definitions of IaIs of the kind that are used when they are audited.

Using this book

There are enormous variations in the quality, quantity and arrangement of the teaching about infections and their control that is offered to students on courses in the healthcare sciences. No single book can act as a primer for all. The transfer of any information may be likened to a journey down a river, sometimes rushing over rapids, sometimes passing slowly through reflective meadows. Biological ideas do not flow easily if they are written (though they may be taught) in a succession of soundbites, like strings of sausages. This is why we have not attempted to conform to any particular teaching format, nor do we provide information in packages that might fit into standard time-tables. The book is not a set of lecture notes. It is the duty of course tutors to dissect the information available according to the time allocated for its delivery and the needs of the class concerned. We attempt to lead readers from topic to topic to complement classwork rather than to act as a model for it. At the same time we have tried to provide a convenient source of reference for individuals who need a deeper insight into the subject of infection control.

Iatrogenic infections

Old and new ideas about infections – what they are, what causes them, how they spread, what has been done to prevent them and the lingering influence of the old upon the new. The dramatic changes that followed the introduction of the antimicrobial drugs and how these sometimes cause harm, particularly in hospitals.

Introduction: contagion and miasma

Not only are infections painful facts of life, they are also the chief causes of death. Even in those parts of the world where major infections have nearly disappeared and cardiovascular disease and cancer are now more common, in the end it is often an infection that tips morbidity into mortality. Fear of infection is deeply rooted in the human mind.

At one time it was believed that diseases were supernatural visitations as punishments for sin. When ancient philosophers discarded this idea they began to separate infectious from non-infectious diseases (a process that is still incomplete). They also developed two opinions of how infections are spread. Some argued this was by contact between individuals, either directly or indirectly by way of inanimate objects passed between them. The process was called contagion and a logical development of the theory required the existence of a transmissible particle that was 'infectious'. Others believed that infection resulted from the inhalation of toxic emanations from the earth. These 'miasmas' were often accompanied by smells, so the foul odour of putrefaction was dangerous. To prevent contagion it was customary to isolate those afflicted, who were identified as sources of infection. In the case of leprosy this was for life, otherwise it might be for 40 days, as the word 'quarantine' suggests. Miasmas, on the other hand, were avoided by flight from the supposed origin of the toxic emanation or they might be removed by clearing away or neutralizing the sources of smells. In the absence of experimental observations, illogical arguments between miasmatists and contagionists continued until the middle of the 19th century.

In the 18th century the contagious nature of gonorrhoea was self-evident to Samuel Johnson's rather promiscuous biographer James Boswell, but how cholera, typhoid and tuberculosis were spread by contact was not clear. Until the part played by blood-sucking insects and other arthropods had been discovered, miasma was as good an explanation as any for the otherwise unaccountable transmission of malaria, plague or yellow fever. The argument lingers to this day in the doubts some have about the relative importance of personal (contagious) and environmental (miasmic) factors as the causes of the infections that may result from medical treatment.

In ancient times hospitals were places to which soldiers, slaves and later the poor and homeless were sent when they were ill, injured or insane. They functioned to keep the sick out of sight rather than to cure them. With the rise of religious orders this tradition was overlaid by more humanitarian ideas, though traces of it persisted well into the 20th century, particularly in connection with the treatment of mental illness. Until very recently the sick who could afford it were treated at home. This was prudent. In the Middle Ages typhus and relapsing fever spread death equally in hospital ward and prison cell and until more recently hospital gangrene and puerperal sepsis exacted a heavy toll. In 1856, the overall death rate following childbirth in the Paris Maternité is said to have been 6%. As a contribution to an argument about the proper siting of hospitals, Simpson (Box 2.1) recorded that, in the preceding decade in England, Scotland and Wales, nearly a quarter of 7264 amputees died shortly after surgery (Simpson, 1869). Most of these deaths were due to infections.

Box 2.1 Sir James Young Simpson (1811–1870)

James Simpson studied medicine at Edinburgh University where, later, he was appointed to the chair in obstetrics. In 1847, after a series of hazardous experiments on himself and his colleagues, he published the first account of the use of chloroform as an anaesthetic. Together with nitrous oxide (Sir Humphrey Davy, 1800) and ether (Dr Morton, 1846), chloroform was used to alleviate pain in dentistry, surgery and in childbirth, in the latter case by Queen Victoria herself. The introduction of general anaesthesia allowed surgery to grow from the performance of a small number of operations applied in emergencies into an acceptable and considered form of treatment for a large range of previously untreatable conditions.

James Simpson was created a baronet in 1866.

Ignaz Philipp Semmelweis (1818–1865)

It is usual (though not necessarily accurate) to date the first scientific attempt to control infections in hospitals to 1844 (Semmelweis, 1861). In that year the Hungarian Ignaz Semmelweis was appointed to the Obstetric Division of the Vienna General Hospital. This had two clinics: medical students were taught in the first and midwives were trained in the second. It was common knowledge that the maternal death rate was significantly higher in the first clinic. Semmelweis set out to investigate this and he began what is now regarded as a classic study by measuring the size of the difference. Although it varied from time to time, the average mortality in the six years from 1840 was 10% in the first clinic and 3% in the second. Most of these deaths were due to puerperal fever, an infection that originated in the wound left in the uterus after separation of the placenta. Semmelweis set out to find the cause of the infection and how to prevent it. He was helped by the existence of the two clinics. One after another, he altered methods and practices in the first clinic and waited to see if this had any effect on the mortality there, compared with the second. Semmelweis may have been the first epidemiologist to use 'test' and 'control' groups in this way. He also made the changes one at a time, to avoid confusing their effects. The experiment took a long time and cost many more lives, but Semmelweis eventually found the cause of the excess mortality and discovered how to reduce it. Bacteriological insight might have made his task easier, but this was not to be available for another 30 or 40 years.

The medical students were taught the elements of obstetrics on the bodies of women who had died after childbirth, many of them from puerperal fever. Staff and students moved freely from post-mortem room to bedside. By 1847 Semmelweis had deduced that a contagious substance was spread from the dead to the living on their hands. In his control group midwives were instructed using models so did not visit the post-mortem room. He made staff and students in the first clinic wash their hands with a disinfectant (a solution of chlorine) after they left the post-mortem room, before they tended their patients. The mortality in the first clinic immediately fell to match that in the second. When he introduced handwashing between the examination of successive patients, mortality fell to less than 1% throughout the whole division.

Unfortunately, Semmelweis was a poor communicator. As a 'colonial' he was also looked upon as an interloper in the great and glamorous capital of the Austro-Hungarian empire. When his ideas were challenged he turned on his colleagues and called them murderers. Not unaturally they were antagonized and they forced him to leave. His work was ignored and a century passed before the importance of the hands in the transmission of infections in hospitals was rediscovered. Another important deduction was missed. It is certain that Semmelweis' patients were dying of classic puerperal fever caused by *Streptococcus pyogenes*. The walls, floors, bedding, furniture and the air in his clinics must have been heavily contaminated with streptococci, yet handwashing on its own had a most dramatic effect. The conclusion ought to have been that the environment, including the air, could at most have been responsible for only a part of the 1% mortality that remained after the hands had been dealt with.

Semmelweis' unfortunate personality ensured that this strong argument against the importance of the environment as a source of an infectious miasma fell into obscurity with his other discovery.

Sewers and drinking water

During the early part of the 19th century, for the first time since the Roman era, many of the great cities of the world were equipped with efficient sewers and their citizens began to receive supplies of clean water. As open drains and cesspits vanished so did the smells and the separation of sewage from water used for drinking and the preparation of food led to a marked reduction in the incidence of cholera, typhoid fever and other enteric infections. The simultaneous disappearance of the smells and of these diseases was linked as cause and effect, once more to encourage a belief in miasma as the cause of infection.

Pasteur and Lister

Paradoxically, the birth of the science of microbiology also lent support to the miasmatists' belief that infections are spread through the air. Louis Pasteur's interest in fermentation led him into a dispute with those who believed in the spontaneous generation of life (Box 2.2). Pasteur showed that living things appeared in nutrient fluids only after they had been exposed to the air, so that small airborne particles could fall into them. By 1864 he had disposed of the theory of spontaneous generation, but at the same time he had showed that air

---

Box 2.2 Louis Pasteur (1822–1895)

Louis Pasteur was the son of a sergeant-major who served in the French Army under Napoleon in the Peninsular War (1809–1814). He started his professional career as an organic chemist, but he soon became the first microbiologist.

An industrialist who produced alcohol from sugar beet found that some of the vats in which the reaction was carried out produced a useless acid instead. Pasteur, asked to investigate, found that the 'sick' vats contained large numbers of a certain bacterium, while the 'healthy' ones contained a yeast. He deduced that the acid and the alcohol were produced from the sugar by the bacterium and the yeast, respectively. He found that the unwanted bacterium could be killed by heating the beet juice before the yeast was added. He published the first account of the activities of microbes in 1857 and the heating process was subsequently called pasteurization, in his honour.

Later, Pasteur disposed of the theory of the spontaneous generation of life, and discovered the cause, and devised a cure, for an infection of silkworms that had threatened to destroy the French silk industry. He and his team produced vaccines against fowl cholera, anthrax, and swine erysipelas, and against human rabies.

In the course of his work he developed many microbiological techniques that are still used today. Although most of his achievements lay in the industrial and veterinary fields, he is justly honoured as the father of medical as well as of general microbiology.

contains the 'germs' of putrefaction. The fact that these are present in small numbers and are rarely of the kinds that cause human disease did not appear until much later. Pasteur's experiment was impressive and easily repeated. The surgeon Joseph Lister (Box 2.3) did so and concluded: '. . . putrefaction in surgical practice is due to particles of dust ever present in the atmosphere'. For him putrefaction was synonymous with infection. He attacked germs in wounds with a carbolic acid paste and particles of dust in the air of the operating room with his famous carbolic spray. Because John Tyndall had shown that cotton wool could filter particles out of the air, he used this as a dressing to protect wounds postoperatively. Lister introduced the elements of antisepsis into his surgical practice in the latter half of the 1860s and quickly achieved an impressive reduction in postoperative infection (Lister, 1871). In fact, his success had more to do with the general disinfectant action of carbolic acid, which in the case of the spray settled on otherwise unsterile hands, instruments and, perhaps particularly, on ligatures and sutures. The spray had little if any effect on the air. Lister eventually realized this and publicly renounced his former belief at the Tenth International Medical Congress in Berlin: '. . . I feel ashamed that I should have ever recommended it' (Lister, 1890). His recantation excited little notice, however, as by the time he made it 'aseptic' surgery had almost completely replaced the 'antiseptic' variety.

Box 2.3 Lord Joseph Lister (1827–1912)

Joseph Lister was the second son of a successful Quaker wine merchant. In 1843 he entered University College, London, to study medicine. He was the first of his family to go to university. Dissenters were excluded from Oxford and Cambridge and in England there was nowhere else to go until the non-sectarian London University was founded in 1828. In 1846, at University College Hospital, he witnessed the first use in England of ether as an anaesthetic. He became a Fellow of the Royal College of Surgeons in 1852 and in 1853 he moved to Edinburgh as assistant to the famous surgeon James Syme, whose daughter Agnes he married in 1856. In the latter half of the 1860s he began to use carbolic acid on dressings and to spray it into the air in his operating room. Surgeons, encouraged by the useful reduction in the incidence of postoperative sepsis produced by this 'antiseptic' surgery, were able to develop more adventurous techniques and so exploited the benefits of anaesthesia more fully than would otherwise have been possible.

Lister was created a baron in 1897, the year of Queen Victoria's diamond jubilee.

Florence Nightingale

Not everyone accepted the implications of the new science of microbiology. Florence Nightingale's passionate faith in the value of fresh air and cleanliness in the prevention of diseases in general and of infections in particular led her to reject them (Box 2.4). A simple

Box 2.4 Florence Nightingale (1820–1910)

Florence Nightingale set out to improve the unhealthy and squalid physical conditions endured by patients in the hospitals of her time and to raise standards among the nurses who cared for them. Her considerable ability as a persuasive administrator, her dogged single-mindedness and her autocratic refusal to accept defeat, together with the help of powerful friends in high places, allowed her to achieve a series of successes that were widely acclaimed in her own lifetime. After the Crimean War the (then enormous) sum of £50 000 was raised by public subscription to mark her activities as 'the lady of the lamp'. She used the money to found the Nightingale Home for the training of nurses at St Thomas's Hospital in London.

Florence Nightingale's belief that disease was caused by foul air and dirt epitomized the view, widely accepted in late Victorian times, that 'cleanliness is next to godliness'. In her *Notes on Nursing* (1859) she wrote: '. . . air (that is) stagnant, musty, and corrupt . . . is ripe to breed small-pox, scarlet fever, diphtheria, or anything else you please'. She was over 60 when the new science of microbiology not only challenged her apparent belief in the spontaneous generation of life, but also began to identify the real causes of infections. She would not accept these revolutionary ideas, scorned what she called 'the germ fetish' and remained a miasmatist to the end of her days.

extension of her ideas led to the conclusion that hospitals in dirty, smelly, crowded towns are more dangerous than those sited in the country. In her *Notes on Hospitals* (1863) she used data collected by William Farr, the British Registrar General, to show that, as measured by their mortality rates, large municipal hospitals were more dangerous than small country hospitals. This led to an argument about where to build hospitals, highlighted at the time by the need to move St Thomas's Hospital from the site it had occupied for six centuries at the southern end of London Bridge.

Florence Nightingale's preference for hospitals in the country was attacked by Sir John Simon, Medical Officer to the Privy Council (the equivalent of today's Chief Medical Officer to the Department of Health). Simon commissioned Dr J. S. Bristowe, physician to St Thomas's Hospital and Mr T. Holmes, surgeon to St George's Hospital and the Hospital for Sick Children, to conduct a survey. They were to '. . . ascertain the influence of sanitary circumstances in different hospitals . . . (on the) more or less successful results for medical and surgical treatment . . . in which recovery is delayed or prevented by accidental morbid complications'. Over a period of nine months, Bristowe and Holmes visited 101 of the larger hospitals in England and Scotland, as well as some in Ireland and Paris. They collected details of their siting, structure and the use made of them and recorded the mortality experienced in each. Their findings and conclusions (Bristowe and Holmes, 1864), together with Simon's comments, appear in the Report of the Medical Officer of the Privy Council for 1863. They concluded that large hospitals were just as safe as smaller ones and added:

'The rules of hospital hygiene are simple, . . . fastidious and universal cleanliness, the same never-ceasing vigilance against the thousand forms in which dirt may disguise itself, in air and soil and water, in walls and floors and ceilings, in dress and bedding and furniture, in pots and pans and pails, in sinks and drains and dustbins.'

Although the position adopted by Nightingale and Farr was supported by the results of Simpson's 1869 survey, the argument went against them and the new St Thomas's Hospital was built opposite the Houses of Parliament at the eastern end of Westminster Bridge, where it still stands.

Despite this Florence Nightingale and her supporters did have some success. In 1871 the press of the time noted the pavilion style of architecture used in the rebuilt St Thomas's Hospital. This was praised because it '. . . prevented the congregation of too many patients under one roof for . . . it is now well known that . . . walls, ceilings and floors become saturated with mephitic odours . . . (that) give rise to those terrible after-consequences of operations known as hospital gangrene'.

A fierce, sometimes acrimonious controversy did not resolve the argument about the siting of hospitals. The major reason for the difference was that hospitals in overcrowded towns and cities treated large numbers of patients who had suffered severe compound fractures of limbs as a result of accidents with unprotected machinery at work or the iron-shod wheels of carts in congested streets. Another reason was that the data used by the two sides had been collected and analyzed in different ways. The fact that the discussion took place at all, however, and the publicity that surrounded it was useful. Since that time everything about hospitals, from the way they are built to how they are cleaned, reflects some thought (even if misguided) about the prevention of infection. Many of the errors that have been made arise because people often fail to distinguish between 'cleanliness' of proven value in the control of infection and 'cleanliness' of aesthetic or imaginary importance.

Streptococcal infections in hospitals

The streptococcus was not discovered until 1877, but it was quickly identified as an important pathogen and recognized as a cause of the most serious kinds of infections in hospitals. By then the situation had improved since the days of Semmelweis and it continued to do so. By the 1920s in the UK the maternal death rate from puerperal sepsis had fallen to two for each 1000 births. Nothing more happened until, in the 1930s, bacteriologists learnt to distinguish between different types of *Strep. pyogenes*. The new typing scheme showed that streptococcal infections usually arose from patients' attendants and not from patients themselves. (Semmelweis had discovered this 90 years earlier.)

Bacteriologists also found that they could recover streptococci from the environment surrounding people who carried or were infected with streptococci. This led to the conclusion that the environment is an important source of infection and without the rigorous investigation the situation required, the observation was used as a 'scientific' reason for the introduction of a number of control measures.

These measures included the isolation of infected or carrier patients, the regular swabbing of staff to locate carriers (who were excluded from work until they were cleared) and, so far as the environment was concerned, the practice of damp dusting, the sterilization of bedding, the use of filtered air in special environments like operating rooms, the increased use of disinfectants and antiseptics and so on (Williams *et al.*, 1966). These measures were introduced in a blunderbuss fashion so their individual effectiveness was not, nor could be, measured. Semmelweis would have been puzzled by the need for some of them and he surely would have disapproved of the lack of controlled studies to show that they worked. The introduction of the sulphonamides in 1936 and of penicillin a little later further confused the picture. By 1946 the mortality from puerperal fever in the UK had fallen to three for each 10 000 births. Much of the credit for this final fall goes to the antimicrobial drugs. It is difficult for those who did not experience it to appreciate the magnitude of the revolution in medical practice that followed the introduction of the antimicrobials. An insight is provided by Clive Butler, whose surgical career spanned the critical period (Butler, 1979).

Crucially, though too late to influence events, it was shown that streptococci exposed to the air after they have been shed from the body of a person with an infection suffer from progressive damage before they die. In their premorbid state they can still be detected by growing them in the laboratory, but they have lost their virulence. Other than in special circumstances, they can no longer cause infections (p.50). The validity of many of the measures designed to 'control' streptococci was challenged by this observation, but it was made after the introduction of penicillin had made the precautions irrelevant, so it was ignored. This was a pity, because from time to time essentially the same unproven measures have been reintroduced to deal with outbreaks of infection caused by different microbes (including staphylococci), in quite different circumstances.

A feature of microbiology that is both a strength, and, as was shown with the streptococci, also a weakness is an ability to detect biological material in minute quantities. For most of the time since the science was founded nearly 150 years ago, microbiologists have used methods of much greater sensitivity than were available to other scientists. The weakness of this is that microbes damaged to the point that they have lost their virulence, or in numbers many times fewer than are required to initiate an infection, are readily detected in the laboratory. To this day the discovery of microbes in the environment, of quite unproven relevance to infection, can attract unwarranted attention and lead to inappropriate activity. Those concerned then tend to regard the matter as closed and look no further for the real cause of the problem.

Right at the end of the period of streptococcal dominance there was an important development in the treatment of surgical wounds. In the 1930s and 1940s the treatment of wartime injuries was a major problem and the technique of delayed primary suture (DPS) was very properly reintroduced to prevent gas gangrene. DPS converts a wound into something resembling a third-degree burn. Wounds treated by DPS, burns, other chronic ulcers and the infant umbilical stump all share the same outstanding potential for colonization. Small numbers of bacteria that in most other situations are non-infectious or bacteria of low virulence or that have lost their virulence are able to settle and multiply in the absence of defence mechanisms that would exclude them from other parts of the body. Once such a colonization has been established there is a high probability that the now very much larger population of fully virulent microbes will spread into the surrounding living tissue to cause an infection. In one series of war wounds the incidence of streptococcal colonizations or infections rose from 8% at the time of operation to 30% 14 days later, while those due to staphylococci doubled from a starting point of about 40% (Miles, 1944). The deduction was inescapable: these *open* wounds had been infected post-operatively, in the ward. These findings reinforced earlier perceptions and the 'no-touch' dressing technique was introduced to solve the problem. Both the idea and the new technique persisted. This is illogical because most peacetime surgical wounds are closed immediately and after a few hours are impervious to microbial invasion from the outside (see p.141 for more details).

Staphylococcal infections in hospitals

When, in the mid 1940s, the streptococcus ceased to be an important hospital pathogen, it was replaced by the staphylococcus. Although infections due to staphylococci had always been more common, the lower virulence of this organism explains why infections due to it were not taken seriously while the much more pathogenic streptococci were more prevalent. Unmasked by the virtual disappearance of the streptococcus, *Staphylococcus aureus* soon became resistant to penicillin and so consolidated its position. This development was the result of the early widespread and, with the benefit of hindsight, quite irrational use of the drug after it was introduced in 1941. First in hospitals and then in the community, resistant strains appeared in such numbers that they virtually replaced the sensitive kind.

The emergence of this resistance depended on the prior existence of a few staphylococci that produced an enzyme able to attack and destroy penicillin. This enzyme (first called penicillinase but later renamed beta-lactamase) inactivates the drug before it can reach and harm the bacterium that secretes it. For millennia staphylococci have been exposed to small quantities of the natural product of various *Penicillium* fungi, one of which was eventually used to make the penicillin 'invented' by humans. Bacteria have had many centuries to develop an effective defence against this form of attack. In the last 50 years they have been exposed to large quantities of the human product. The offspring of the few staphylococci that were naturally resistant to the drug because they produced a beta-lactamase soon filled the gap left by their sensitive relatives. The result was that staphylococcal infections no longer responded to treatment with penicillin. Staphylococci were soon recognized and came to be feared as the most important causes of infections in hospitals.

The Gram-negative rods

This return to preantibiotic therapeutic impotence in the face of a common form of infection lasted until 1960 when methicillin, the first of a series of penicillins resistant to staphylococcal beta-lactamase, was introduced. With the staphylococcus once more under control the role of dominant hospital pathogen passed to a succession of Gram-negative rod-shaped bacteria (GNRs), at first notably klebsiellas and serratias. These were responsible for relatively small outbreaks of infections in hospitals which attracted attention because they were readily identified and because little else was happening at the time. The infections they caused were more difficult to treat because the GNRs responsible had, like the staphylococci before them, become resistant to antimicrobials. This time the resistance was to the broad-spectrum drugs that came into general use in the 1960s. Ampicillin was an early example.

Some GNRs continued the process and have become increasingly and multiply resistant to antimicrobials. They are still important causes of infections in hospitals, but back in 1976 they lost their dominant position when a new breed of staphylococcus appeared. These were resistant not only to penicillin but also to methicillin and to the whole range of the newer penicillins and cephalosporins, as well as to the important broad-spectrum aminoglycoside antimicrobial, gentamicin. At the time of writing, in many parts of the world, colonizations and infections with methicillin- (and often aminoglycoside-) resistant *Staph. aureus* (MRSA) are widespread, not only in hospitals but also in the community. In some places MRSA attracts much anxious attention and expensive reaction. Elsewhere it is ignored. There is no persuasive evidence that either approach is superior to the other (p.165).

Antimicrobial resistance

Each time a new antimicrobial drug is introduced, some microbes that were at first sensitive to it become resistant. Such organisms tend to cause serious infections in hospitals. The constant fear has been and still is that we will 'run out' of effective treatments. As noted, this actually happened with the staphylococcus in the 1950s and the threat of it has been repeated in the 1970s, 1980s and 1990s. In addition to the GNRs, significant levels of antimicrobial resistance are now found in some streptococci (notably in the important pathogen *Strep. pneumoniae* and more recently among the enterococci), and in *Mycobacterium tuberculosis*. Effective antiviral antimicrobial drugs appeared more recently, but a similar pattern has begun to emerge as some viruses have developed resistance to them. The same thing has happened among the fungi.

Most of the bacteria that have developed important levels of antimicrobial resistance belong to species widely distributed among the human population as part of their normal flora. Many are only indifferently pathogenic. It is notable that resistant strains tend to emerge in hospitals. The reason for this is twofold. First, individuals with reduced resistance to infections due to illness are concentrated in hospitals and second, at any one time, not less than one-third of them are receiving antimicrobial drugs. In these circumstances relatively non-pathogenic microbes tend to act as full pathogens which, simultaneously, are exposed to a selective pressure that encourages the emergence of resistance to the antimicrobials in use. Hospitals are first-class seedbeds in which to grow multiply resistant strains of the more common bacteria.

The observation that resistant strains of bacteria *appear* in hospitals has been extended to include the erroneous assumption that hospitals *caused* the resistant mutants to emerge in the first place. Small numbers of (at that time) undetectable beta-lactamase-producing staphylococci existed before penicillin was introduced, but they were not selected to emerge as a serious problem until the late 1940s when the drug was used on a large scale. Hospitals are not microbiologically watertight. When a patient who has been colonized or infected with a resistant microbe in a hospital goes home, the resistant bacteria tend to be replaced by more sensitive ones, though this happens less quickly in infants and the elderly. At some point the resistant strain seems to 'disappear' as their number falls below the sensitivity of the laboratory method used to detect it. It is a biological certainty that it takes much longer for the last resistant bacterium finally to disappear, if indeed it ever does. The population is continuously and eventually permanently 'seeded' with small numbers of resistant strains of bacteria. These stay out of sight until they re-emerge (that is, increase in numbers until the laboratory can detect them once more) when the patients who carry them are exposed to appropriate antimicrobial drugs.

A further source of this 'seeding' comes with the food we eat. The intensive production of animals and birds for human food requires that they consume quantities of powerful antimicrobial drugs. This is necessary to prevent or suppress infections that would otherwise kill them or at least unacceptably lengthen the time for which they must be fed to reach a marketable weight. Modern animal husbandry is even more effective as a seedbed for resistant bacteria than are human hospitals. It is irrelevant that the antimicrobials used are supposedly (though not always) different from those employed in human medicine. The genetic mechanisms that control antimicrobial resistance in bacteria operate so that exposure to one antimicrobial can encourage the proliferation of resistance to others, even when the drugs are of different kinds. Resistant bacteria that re-emerge when those who carry them are exposed to an appropriate antimicrobial are more likely to appear first in hospitals, so the bacteria responsible (not always with justification) are called 'hospital strains'. A final implication of the widespread, low-level seeding of resistant microbes through a community is that the routine swabbing of patients and staff in an attempts to control their spread is, in the end, bound to fail.

It is fortunate that *Strep. pyogenes* has not yet followed the example of some of its relatives that, between them, have developed resistance to penicillin and to many other important antimicrobial drugs. If *Strep. pyogenes* were to acquire these resistances the stench of 'hospital gangrene' would once more return to our hospitals, and, as in the days of 'laudable pus', an infection with *Staph. aureus* might be welcomed as a desirable, because less lethal, alternative.

REFERENCES

Bristowe, J. S. and Holmes, T. (1864) Report on the hospitals of the United Kingdom, in *Sixth Report of the Medical Officer of the Privy Council, 1863*, Her Majesty's Stationery Office, London, pp. 463–743.

Butler, C. (1979) Surgery – before and after penicillin. *British Medical Journal*, **ii**, 482–3.

Lister, J. (1871) The address in surgery. *British Medical Journal*, **ii**, 225–33.

Lister, J. (1890) The present position of antiseptic surgery. *British Medical Journal*, **ii**, 377–9.

Miles, A. A. (1944) Epidemiology of wound infections. *Lancet*, **ii**, 809–14.

Nightingale, F. (1859) *Notes on Nursing; what it is, and what it is not*. Reprinted in facsimile 1974, Blackie, Glasgow, p. 8.

Nightingale, F. (1863) *Notes on Hospitals*, 3rd edn, Longman, London, p. 3.

Semmelweis, I. P. (1861) *Die Aetiologie, der Begriff, und die Prophylaxis des Kindbettfiebers*, Pest, Wein u. Leipzig.

Simpson, J. Y. (1869) Some comparisons, etc. between limb amputations in the country practices and in the practices of large metropolitan hospitals. *Edinburgh Medical Journal*, **15**, 523–32.

Williams, R. E. O., Blowers, R., Garrod, L. P. *et al.* (1966) *Hospital Infection*, 2nd edn, Lloyd-Luke, London, pp. 9–21.

Iatrogenic infections: basic facts

COLONIZATIONS AND INFECTIONS

The relationships that develop between the human body and the teeming populations of microbes that live on and around it are examined. Possible outcomes for microbes vary between exclusion, through the establishment of colonizations, to the development of infections. When normal colonizations are upset disease may result. The distribution of the 'normal flora' and its importance in the maintenance of health are discussed. Epidemic and endemic infections are defined.

Introduction

Colonizations and infections are caused by **microbes**, that is by viruses, bacteria, algae, protozoa, fungi or by members of a group of very small infectious agents that have been called viroids or **prions**. These differ from each other and from animals and plants in ways described in microbiology books. Viruses and prions excepted, microbes possess all the characteristics of living creatures. Viruses are very small with space for little more than the instructions they need to duplicate themselves. Mature virus particles are not really alive – their potential for it remains unfulfilled until they enter and multiply inside the cells of other living things. Microbes are by far the most numerous living things on earth (Box 3.1). Most are beneficial and animals, plants and the human race itself could not exist without them. A few of them have learnt to attack other living things, to cause **infections**. Larger parasites, such as worms, are not microbes and the diseases they cause may be called infestations rather than infections.

Microbe A living organism (or a virus or prion) that is too small to be seen with the naked eye.

Prion 'Proteinaceous infectious particle': a self-replicating subviral body uniquely free of nucleic acid.

Infection The outcome of an interaction between a host and a microbe in which the host reacts in some observable way.

The surface of the body

From shortly after birth to beyond the grave, every human being is host to enormous numbers of microbes. Among these microbes bacteria attract attention because they are responsible for a large proportion of the infections that develop in or after admission to hospitals or as a result of medical care delivered elsewhere (collectively **iatrogenic infections** (IaIs)). Viruses and fungi are next in importance.

Iatrogenic infection An infection that is the consequence of a medical intervention.

Box 3.1 From animalcules to microbes

The world is thought to be 4.5 billion years old. Microbes seem to have appeared on or near the surface of it 3.5 billion years ago, so they are the oldest as well as the most numerous living things on earth. Microbes ('germs') are invisible to the naked eye, so are less than one-tenth of a millimetre (100 μl) across. The first person to see them was Antonie van Leeuwenhoek (1632–1723), a Dutch draper and haberdasher and amateur lens maker. In 1665 he constructed a magnifying glass that allowed him to see the microbes that had grown in a tub of stagnant water. The pursuit of what he called his 'animalcules' became a lifelong hobby. His correspondence with the Royal Society in London and the Académie des Sciences in Paris excited much philosophical interest, but the deep significance of his discovery was not recognized until the science of microbiology was founded by Louis Pasteur 200 years later.

Normal flora A collection of microbes, usually present as colonists, in any specific environment, including some surfaces of the healthy human body (on the **colonized surfaces**).

Colonization A long-term relationship in which a microbial population lives on (or in) a host, with no observable reaction by the host to the presence of the microbes.

Bacteria inhabit some surfaces of the body, where they form a '**normal flora**'. Each of us carries a normal flora of between 10^{14} and 10^{15} bacteria (one followed by 14 or 15 zeros or about one for every millimetre that separates the earth from the sun). This huge number compares with a paltry 10000 to be found on a square metre of somewhat dirty floor or 1000 in a cubic metre of the air in an occupied, poorly ventilated room. If it were possible to make such a distribution, an individual could hand out 100000 members of their normal bacterial flora to every person on earth and still have at least 100000 left. In about 24 hours the residual 100000 would have multiplied to reinstate the former population. This extraordinary reproductive ability explains why bacteria are so important as causes of IaIs and why many infections are caused by bacteria that were already present on an individual before it became apparent.

Depicted from the viewpoint of a bacterium, the human body is divided into two zones (Figure 3.1). The surfaces of the body that are normally inhabited by bacteria make up the first zone, represented by the outer ring in the figure. These are the **colonized surfaces** of the body. The inside of the circle is the **forbidden zone** which, in healthy people, is kept virtually free of microbes. This is divided into two parts, **privileged surfaces** and **internal tissues**. Much of what is conventionally thought of as 'inside' the body is really a part of its surface. The lungs, gastrointestinal and genitourinary tracts, for example, could not perform their biological functions if they were shut off from the outside world. The privileged surfaces have an area many times greater than that of the visible skin but they are conveniently folded away, out of sight. It is important to recognize these as true surfaces, however, because many of the infections that arise in hospitals begin when microbes present on a normally colonized surface spread to invade the internal tissues, in many cases by way of one of the privileged surfaces (Box 3.2).

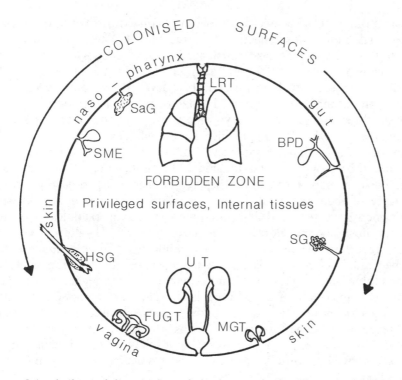

Figure 3.1 A 'bacterial eye' plan of the human body. The colonized, easily accessible surfaces (skin, gastrointestinal tract, etc.) are on the outside of the circle and the 'forbidden zone', that, in health, is kept substantially free of microbes, is in the centre. The forbidden zone is subdivided into the 'privileged surfaces' and the 'internal tissues'. Although all privileged surfaces are in direct structural continuity with one of the colonized surfaces, they possess special mechanisms that prevent the microbes present on the latter from spreading onto the former.

Key: LRT, lower respiratory tract; BPD, bile and pancreatic ducts; SG, sweat glands and the breasts; MGT, FUGT, male, and female upper genital tracts; UT, urinary tract; HSG, hair follicles and sebaceous glands; SME, sinuses and middle ear; SaG, salivary glands.

Box 3.2 From a colonization to an infection, via a catheter

A urinary catheter does more than relieve a problem with the flow of urine (p.130). In health the distal or terminal part of the urethra is colonized by bacteria from the surrounding skin. With a self-retaining catheter in place this colonization is encouraged to spread backwards in the space between it and the mucous membrane that lines the urethra. Bacteria that use this route eventually reach the bladder, where they multiply vigorously in the warm, wet environment. If the catheter is not removed in time, the bladder wall eventually responds to the insult, becomes inflamed, and an iatrogenic urinary tract infection is established. The infection may extend further to reach the kidneys, from where bacteria can more easily enter the bloodstream, to cause a second but now potentially lethal infection, iatrogenic septicaemia.

The outcome of a meeting between a collection of microbes and a potential human host depends on three things. These are the relative strengths of, first, the **host resistance** (the effectiveness of the defence mechanisms they deploy to counter the invasion), second, the **virulence** of the attacking microbe (the **level of its pathogenicity**) and third, the number of invaders (the **size of the challenge dose**). Each of these three factors can vary widely so, in any given case, the outcome is difficult to predict. From day to day, unimaginable numbers of microbes come into contact with the surfaces of our bodies. Most of them fail to gain a foothold. Very few manage to establish a permanent bridgehead and begin to multiply on the surface of their new host. If the host does not react to this invasion, the microbe is regarded as a **non-pathogen** and it becomes a member of the host's **normal flora**. This is a colonization. If, alternatively, the host reacts to the invader in a detectable way an **infection** is established and the microbe is now a **pathogen**. If the reaction is readily observable (as nausea, vomiting, diarrhoea, headache, fever, rash, muscle pains, etc.) the individual is ill. They are then said to suffer from a **clinical infection** as the body signals its discomfort ('dis-ease') in reaction to some change in the distribution of microbes on it or in it. Otherwise the reaction is so inconspicuous that it can only be detected by a laboratory test. This is a **subclinical** or **asymptomatic infection**.

Pathogen A microbe or larger parasite that can cause harm to a host, to produce disease (an infection or infestation). **Pathogenicity**, the expression of this property.

Infection: host defences

Infections are unique among diseases because they involve two quite distinct living systems, the microbe and its host. To understand the process of infection it is necessary to know something about each of the parties involved and how they interact. On the human side there is considerable variation in the vigour of an individual's response to microbial attack. The normal human body has a remarkable array of defences against microbes. This **immune system** is incompletely developed in the very young or it may be permanently defective due to an inborn error. It is temporarily weakened in pregnancy and parts of it wear out as we get older. It is disturbed by illness (particularly malignant disease) and various medical or surgical treatments may inactivate or bypass important parts of it. Any or all of these causes of **immune deficiency** lower individuals' resistance to infection, called **immunocompromise** when it is temporary or **immunodeficiency** when permanent. The result of a deliberate attack on immunity is **immunosuppression**.

The body's defences are of two kinds. Those of the first kind are permanent and are present from birth or develop soon after it. They form the **innate immune system** that is made up of non-specific **physical, chemical** and **cellular** barriers that act unselectively against most microbes. The second set comprise the **adaptive immune system**. This responds to an infection by producing defences (of which antibodies are an example) that are selectively tailored to attack the microbe responsible. They act to terminate the infection and then persist to prevent another invasion by the same microbe. The innate system provides the principal barrier to what otherwise would be an inevitable invasion of parts of the forbidden zone by microbes from neighbouring colonized surfaces (Figure 3.1 and Box 3.2). For example, they act to ensure that the microbes that form the normal flora of the throat normally do not spread into and colonize, or infect, the lungs.

A feature of medical treatment is the frequency with which these natural defensive barriers are bypassed by such things as surgical wounds, urinary catheters, endotracheal tubes and various intravascular devices. Each of these devices provides a highway along which microbes can penetrate the forbidden zone to cause colonizations and infections in parts of the body otherwise inaccessible to them. Infections that arise as a direct consequence of these or other forms of medical or surgical intervention are, of course, **iatrogenic infections**.

Infection: microbial virulence

A complete spectrum of degrees of virulence lies between **non-pathogenic commensals** and the most pathogenic kinds of microbes. **Fully virulent pathogens** may cause infections in healthy people. To do so they must be introduced in sufficient numbers by the correct route and the individual must not already be immune due to a previous infection with the same pathogen or as the result of a vaccination. The measles virus is a fully virulent pathogen because it nearly always causes disease when encountered in its wild (natural) form. The individual is then immune to a further attack by the measles virus. On the other hand, if the first meeting is with the heavily modified virus used in measles vaccine, the result is almost always an asymptomatic infection though once more the immunity that develops prevents a further attack by the measles virus. The intermediate part of the spectrum of virulence is occupied by microbes that are labelled **potential** or **opportunistic pathogens**.

The situation is entirely different in patients suffering from an immune deficiency. Such patients can be infected by weak pathogens of low virulence, even when they attack in small numbers. Because their immune systems fail to respond properly, unusually prolonged or repeated infections are possible, which may be caused by opportunistic pathogens or, in extreme circumstances, by commensals. For such an individual the line that separates pathogens from non-pathogens has disappeared.

The other important variable is the number of microbes that make the attack, that is, the **size of the challenge dose**. The defences of the body can repel surprisingly large numbers of bacteria. For example, it is necessary to inject a million *Staphylococcus aureus* into healthy skin to produce a small pustule. Ten million painted onto intact skin have no effect. The situation is different if immunity is compromised. One hundred *Staph. aureus* on a suture will produce a stitch abscess. The presence of this small foreign body is sufficient to reduce immune competence by a factor of 10 000 (Elek and Conen, 1957).

The normal flora

The normal flora of the body is complex, with a powerful built-in resistance to change. Bacteria from the environment that attempt to take up residence rarely succeed. This resistance to colonization depends in part on the existence of a balanced normal flora. If the balance is upset an existing component may overgrow and new colonists are able to establish themselves much more easily (Box 3.3). Common events in medical practice, such as treatment with antimicrobial drugs, exposure to antiseptics, changes of diet or therapeutic starvation, can all upset the balance. Many patients add new colonists to their 'normal' floras shortly after they come under medical care. In this case the microbial newcomers are likely to have become acclimatized to life in the healthcare environment, so are more resistant to antimicrobial drugs and more easily take advantage of any special circumstances in the unit concerned (Box 3.4). These have been called 'hospital strains' of bacteria. Although they are sometimes called 'superbugs' (p.164), they rarely threaten healthy, immunocompetent people and so are uncommon causes of infections among hospital staffs and their families. For the same reason patients who leave hospital carrying some of them are not

---

Box 3.3 A spoilt holiday?

The 'balance of the normal flora' refers to the state of the mixture, measured by the range of different types of microbe, and the numbers of each, in the whole population (the 'flora') carried by a healthy individual. This balance may be upset, for example, by a move from a temperate climate into a hot and humid one. Many people discover this for themselves the hard way when they travel to a new job in a tropical country or even when they take a holiday in the sun. Itchy rashes ('prickly heat') or the appearance of tinea pedis (athlete's foot) or of tinea cruris (jock itch), perhaps in virulent forms, are the more common manifestations of this. When the same thing happens in the gastrointestinal canal, 'traveller's diarrhoea' is the result.

---

serious threats to their friends and relatives. If and when patients regain health and immune competence any hospital bacteria they still carry tend to be replaced by a more normal and more healthy flora.

The surface of the human body resembles many other biological systems in which a mixture of different living things come to terms

Box 3.4 New colonists in hospitals

Pseudomonas aeruginosa only attacks tissues that are compromised in some way, so it is a good example of a non-pathogen that acquires a pathogenic potential among debilitated patients. Like most iatrogenic pathogens it tends to establish a colonization on one of the body surfaces before it spreads to cause an infection. Such colonizations often involve the gastrointestinal canal. In the community about 20% of individuals carry small numbers of *Ps. aeruginosa* in their intestines. After a few days in an ordinary hospital bed the proportion may rise to 30–40% or to 90% after a week in an intensive care unit and the carriage is no longer scanty. Infections due to *Ps. aeruginosa* are at least twice as common among patients following their admission to hospital than they are elsewhere.

with each other to achieve an ecological balance. Such systems have several things in common. For example, in a 'healthy' primary jungle a large number of different species of trees, plants, animals, birds, insects and so on live together in a stable, balanced state, with each species present in relatively small numbers. If primary jungle is destroyed, it is replaced by an 'unhealthy' secondary jungle. This contains many fewer different species, but with each of them present in larger numbers. For example, rodents may proliferate in secondary jungle. These can carry microbes hazardous to humans, so in these situations diseases like leptospirosis, scrub typhus and plague become more common.

In the ecological sense there are similarities between secondary jungle and what happens in hospitals. In health the surface of the human body carries many different microbial species that live together in a balanced state. In an unhealthy unbalanced state (as in secondary jungle) a smaller number of species is found with one or more of them present in abnormally large numbers. The major difference between primary rainforest and the surface of the human body is that in the jungle the lifespan of many of the species is measured in years or decades. On the body most microbial lifespans are measured in minutes. In a few hours a healthy normal flora can become an unhealthy, abnormal one. Fortunately, the opposite is also true and 'health' is regained in a few days rather than the centuries it takes to remake a primary jungle.

Iatrogenic infections

Infections are inseparable from communal life, but they are more common and more severe in hospitals than elsewhere. Some patients come into hospital for the treatment of established infections and others are more or less debilitated by the illnesses for which they are admitted. Treatment commonly causes further debility. A vicious circle is created in which patients with infections are mixed with those more likely to acquire them and fuel is added as more

susceptible patients are admitted to replace those who are discharged or die. The control of the infections that arise in hospitals
has been likened to digging a well and then trying to keep the water
out!

Antimicrobial drugs are used to treat infections on such a scale
that, in many hospitals at any one time, 30% or more of the patients
are receiving them. They may be given for the treatment of established infections or as prophylactics to try to prevent them or, all to
often, for no good reason at all (p.1). Bacteria that survive in this
environment are selected because they can resist the antimicrobials in
use. This is how **hospital strains** of bacteria emerge to grow vigorously in the fertile seedbed provided for them. This pattern of
therapy and its outcome is repeated, to a lesser extent, in the
community.

Modern medical or surgical treatments almost inevitably cause
some reduction in host resistance to infection. For a patient whose
defences are doubly weakened by their illness and by the effects of
treatment, a commensal becomes an opportunistic pathogen and an
opportunistic pathogen a full pathogen. This deterioration in host
resistance is responsible for most infections that arise in medical care.
The microbes that cause the infections are often members of the
patient's original normal flora or were added to it after treatment
had started.

Septicaemia The presence of
microbes in the blood,
accompanied by symptoms.

Patients with severe immune deficiencies may develop septicaemias
secondarily to quite minor, inconspicuous tissue infections. These
arise from small accidental or deliberate diagnostic or therapeutic
breaks in the epithelial surfaces of the skin, gut, respiratory or
urinary tracts. They may be caused by vascular or urethral catheteri-
zations, endoscopy and so on. The bacteria that take advantage of
these defects vary according to the flora (normal or disordered) of
the site involved. Such patients have often been in hospital for some
time, so they are likely to be colonized by less desirable multiresistant hospital strains of bacteria. This is a common situation in
intensive care units, where patients disadvantaged by severe disability
and the insertion of multiple invasive devices are also more likely to
harbour hospital strains among the flora of their colonized surfaces.
When an infection develops, there is a good chance that a hospital
strain will be responsible.

It is clear that some infections are inescapable consequences of
admission to medical care. There is good evidence that there are
more of these infections than need be though the irreducible
minimum has not been defined (p.85). It may never be, because the
target is moving. As medical science develops, ever more debilitated
patients are subjected to increasingly complex invasive treatments.
Although these treatments may improve the quality of life, the
penalty is that some individuals who are exposed to them suffer an
increase in morbidity and mortality due to the infections that result
from their treatment.

The separation of infections from colonizations

In some circumstances a real practical difficulty arises in making a distinction between infections and colonizations. This happens when, as the result of a culture, potentially pathogenic bacteria are found but it is uncertain if they are present as non-pathogenic commensals (a colonization) or as infecting pathogens. The difficulty is compounded because an infection at any site is preceded by a colonization at the same site, perhaps for several days. In the case of classic infections (measles or chickenpox, for example) the **incubation period** is the time between the arrival of a new microbe on or in the body (after exposure to another case) and the emergence of disease. Its length depends on how long it takes for an initial colonization to be converted into an infection. With measles or chickenpox the change from one state to the other is abrupt and obvious. In other cases, and particularly with iatrogenic infections, the point at which a colonization becomes an infection (if indeed it ever does) may be difficult or impossible to determine.

Examples of this difficulty are seen with pressure sores (bedsores or decubitus ulcers) and burns. Destruction of skin and perhaps of subcutaneous tissues leads to exposure of underlying structures and the establishment of a moist area that is rich in bacterial nutrients, but with a defective blood supply. This not only delays healing but it also limits the body's capacity to react appropriately to any invading microbe. Different kinds of bacteria, including relatively non-pathogenic varieties, take advantage of this situation. Various staphylococci, streptococci and, for example, *Pseudomonas aeruginosa* may be found on the surface of the ulcer. More often than not, the presence of these microbes represents a colonization. A decision to call it an infection requires care, for two reasons. Firstly, this is the point at which antimicrobial drugs may be prescribed. Their use in this situation is unlikely to do more than ensure a continued supply of hospital strains of bacteria. Secondly, the patient is then listed among those who have acquired a iatrogenic infection. The criteria for making the diagnosis of an infection (recognized as particularly difficult in these cases) are the presence or absence of the classic signs of inflammation (pain, swelling, redness, heat and loss of function). When the decision is important (before a skin graft for example) it may rest on a microscopic examination of a biopsy of tissue that straddles the dead and living parts of the ulcer. Other examples of this difficulty will be mentioned later.

Endemic and epidemic infections

When infections in two or more patients are clearly related to each other an **outbreak** or **epidemic** of infection exists, though the latter term is usually reserved for examples involving larger numbers of people. When cases of influenza appear in a community, the existence of an epidemic is fairly obvious. In hospitals the situation is different and it is necessary to be much more rigorous in determining the identity of the microbes isolated from the patients concerned. To show that there is an outbreak of, say, wound infections requires that the bacteria causing each of the individual cases are, so far as it is possible to determine, 'the same' (p.41). This nearly always involves some form of microbiological typing. For example, the species *Staphylococcus aureus* can be divided into many hundreds of distinguishable varieties. It is not enough to note only that a group of infections seem to be related by their microbial causes (broadly defined by species), in time and that the patients concerned are connected by the part of the hospital where they were housed or by the members of staff who treated them. Typing is required in addition to any or all of these features to establish that, so far as can be determined, the microbes isolated are also indistinguishable. If they are not then it is doubtful if an outbreak exists, no matter how persuasive the rest of the evidence. Infections that cannot be linked together as a part of an outbreak are classified as **endemic**. Because most iatrogenic infections are self-infections (p.31), most are also endemic.

THE FREQUENCIES AND TYPES OF INFECTIONS

Both numbers and types of iatrogenic infections are broadly predictable, though an increasing number of those that are *acquired* in hospitals now *appear* in the community. They may be counted in different ways.

Introduction

It has been customary to divide infections into two major classes. Those that appeared after (and were the result of) admission to hospital were described as **hospital-acquired** or **nosocomial infections**: these terms have been regarded as synonymous. Their counterparts were the infections patients brought into hospital with them or, if they were not yet clinically apparent, were acquired in the community and were within their incubation periods at the time of admission. These were called **community-acquired infections**. The abbreviations HAI and CAI are widely used for these two classes of infection. For the reasons set out in Chapter 1 the expression **iatrogenic infection** has been introduced to extend the meaning of HAI in a way that is more appropriate to the delivery of healthcare today. Iatrogenic infection is abbreviated to IaI to avoid the Roman numeral II. As the counterpart to IaI, the term **community infection** (CI) has been adopted to include all infections that are not, by definition, IaIs.

It has not been possible to abandon the former terminology completely. In what follows, HAIs and CAIs appear repeatedly because the data presented were recorded and discussed under this nomenclature. It will help to remember that, when these data were collected, HAI was an even more major component of what may now be called IaI than it is at present.

Definitions

For many years HAI has been defined as:

- an infection found in a patient in hospital that was not present and was not being incubated on admission or, having been acquired in hospital, appeared after discharge.

In Chapter 1 it was suggested that the word 'iatrogenic' best conveys the idea that the types of infection formerly characterized as hospital acquired can also be acquired in other places. The need for a change is heightened by the fact that the proportion of HAIs identified by the 'after discharge' part of the above definition has grown markedly in recent years. An IaI may be defined as:

- an infection that is the consequence of a medical intervention,

where the words 'medical intervention' are interpreted to include the activities of any healthcare worker engaged in their duties. Either definition can be expanded to include infections among healthcare workers that result from their employment.

It used to be relatively simple to distinguish between HAIs and CAIs. In most cases the few difficulties that arose were resolved by a careful examination of medical and nursing records or of the patient. It was rarely necessary to use an arbitrary distinction based on the time of appearance of an infection in relation to hospital admission, though as a last resort an infection that appeared in hospital before the third day might be labelled as CAI and later than that as HAI. Different rules may be applied in prosthetic surgery, in which infections associated with implants may become apparent weeks or months postoperatively.

The simplicity of the new definitions is somewhat deceptive. CIs and IaIs are not always easily distinguished. It will be necessary to separate infections that are part of a disease process (so would have occurred in any case) from those that genuinely result from therapy. When CAIs and HAIs were separated the problem did not arise as the distinction between them depended on an objective geographical location rather than on what, in the case of IaIs, may be a more subjective decision about the cause of an infection.

A residual problem under either definition is the need to distinguish between colonizations and infections (p.24). Doubt can persist after all available information has been taken into account and it is sometimes necessary to defer judgement until the situation clarifies in the days that follow. This difficulty is referred to again in connection with the different types of infection described in Chapter 7.

Frequencies: prevalence and incidence

When CAIs have been counted in hospitals in more developed countries the numerical rates have often been similar to those for HAIs counted at the same time. In less developed communities CAIs tend to be more common and the types of infections that are recorded vary more widely. The composition of HAI, by contrast, is everywhere more constant, though allowance must be made for the types of patients being surveyed. Hospitals with heavy surgical loads will record more surgical wound infections than those where more patients are admitted with medical conditions. Discrepancies also arise because of differences in the length of time the average patient spends in hospital. This varies in different countries, though it seems to be falling almost everywhere. In present circumstances infections that take longer to develop (have longer incubation periods, as with surgical wounds in particular) are increasingly underrepresented. In some cases the 'loss' of HAIs due to this has exceeded 60% (p.148). The loss can be corrected by extending the survey of patients into the community after discharge from hospital, but this is time consuming and expensive. It has not often been attempted. Despite this difficulty, similarities tend to outweigh the differences when the numbers of cases and the types of HAI identified in acute hospitals have been surveyed in various countries. The similarities persist across various systems of healthcare and they seem to be independent of differences in the age and construction of hospitals and of the way they are used. This relative constancy is illustrated in Tables 3.1 and 3.2, where the results of 13 prevalence surveys in nine countries on three continents, over an interval of about 30 years, are presented.

In each of the surveys urinary tract infections (UTIs) were the most common variety of the HAIs recorded, as the cause of between 23% and 45% of all of them. Surgical wound infections (SWIs, 13–26% of all HAIs) were the next most common in eight of the surveys, while lower respiratory infections (LRIs, 10–26% of HAIs) came second in five. 'Other infections' are collections of individually less frequent types of HAI. They range from infections of the skin (including those at sites used for intravascular access) to septicaemia, the most life-threatening variety of HAI. Another of the 'other' infections, gastroenteritis, is less common in acute hospitals in more developed countries, but in some parts of the world it is frequent and severe, particularly among children. It is everywhere more common in long-stay patients, especially among those who need psychiatric care. It is caused by failures in personal and communal hygiene.

If the surveys included in Tables 3.1 and 3.2 are listed in date order, it emerges that the proportion of IaI due to SWI has fallen as time has passed. This may be accounted for by improvements in surgical technique or by the earlier discharge of postoperative patients. As over 60% of cases of SWI may be missed by audits restricted to hospitals (p.148) it is likely that early discharge has had the greater effect.

The studies that produced the figures recorded in Tables 3.1 and 3.2 were all 'snapshot' **prevalence surveys**. An alternative is the longer term **incidence survey**. The techniques differ in ways described in Chapter 4, pp.72 and 73, but it is important to note that they measure different things. In the case of HAIs in particular, the rates determined by one kind of survey must not be compared with those from the other. Comparable populations surveyed by the two methods produce significantly lower figures for the incidence than for the prevalence of HAI (or of IaI).

A large incidence survey (study on the efficacy of nosocomial infection control, 'SENIC', Haley *et al.*, 1985a; Martone *et al.*, 1992) produced a figure for the incidence of HAI in the USA of 5.7% (Table 3.3). This is close to half the median rate measured in the prevalence surveys listed in Table 3.1. The reason for the difference is that patients with infections spend about twice as long in hospital as those who escape them. Though the rates of HAI determined by the two methods must not be compared, this does not apply to the proportions of these totals accounted for by the different types of infections. The proportions found in the SENIC study are given in Table 3.3, together with results from two smaller incidence surveys.

Table 3.1 Rates of hospital-acquired infections as detected by prevalence surveys in a number of countries (dates are those of publication)

Country & date	HAI (%)	Code
1 *Canada*		
1976	8.2	a
2 *Czechoslovakia*		
1988	6.1	b
3 *Hong Kong*		
1987	8.9	c
4 *Scandinavia*		
1978	10.5	e
1980	10.5	f
1980	12.1	g
5 *United Kingdom*		
1981	9.2	i
1996	9.0	j
6 *United States of America*		
1964	13.5	k
1971	5.9	l
1974	15.0	m

Note: Broadly similar criteria were used in these surveys and no obvious confounding factors were present.
For references see Table 3.2.

Table 3.2 The sites of the hospital-acquired infections reported in the prevalence surveys listed and coded in Table 3.1, plus two others

Infection   Percent of all hospital-acquired infections

	a	b	c	d	e	f	g	h	i	j	k	l	m
UTI	39	25	33	31	42	38	45	42	30	23	36	42	33
SWI	24	15	24	13	26	18	21	17	19	11	25	23	17
LRI	26	10	21	18	10	22	12	14	16	23	15	19	23
Other	11	50	22	28	22	22	22	27	35	43	23	16	22

Key: UTI, urinary tract infection; SWI, surgical wound infection; LRI, lower respiratory infection; Other, infections at other sites
Note: The codes and dates given in Table 3.1 for the surveys noted there and in Table 3.2 refer to the following publications: a, Hinton, 1976; b, Sramova, Bartanova and Bolek, 1988; c, French, Cheng and Farrington, 1987; d, Moro *et al.*, 1986; e, Bernander *et al.*, 1978; f,g, Jepsen and Mortensen, 1980; h, Hovig, Lystad and Opsjon, 1981; i, Meers *et al.*, 1981; j, Emmerson *et al.*, 1996; k, Kislak, Eickhoff and Finland, 1964; l, Scheckler *et al.*, 1971; m, McGowan and Finland, 1974.

Table 3.3 The rates and types of hospital-acquired infections in three incidence studies

Incidence of:	*per 100 admissions or discharges*		
	USA (SENIC)	Singapore[1]	UK[2]
All infections	5.7	3.4	9.2
Infected patients	4.5	2.9	7.3
Individual infections	Types of HAI (%)		
Urinary tract	42	41	38
Surgical wound	24	26	23
Lower respiratory	11	13	15
Septicaemia	5	4	5
Other	18	16	19

Notes: 1,2 These studies were partial. (1) lacked the sensitivity to detect all cases of HAI, so gave a lower figure for the rate; (2) excluded certain groups of patients at lower risk of HAI, so produced a higher one.
References: USA, Haley *et al.*, 1985a; Singapore, Meers and Leong, 1990; UK, Glenister *et al.*, 1992.

The figures are comparable to those of the prevalence surveys given in Table 3.2.

A feature of surveys in which cases of HAI (and so of IaI) are counted is that, where the distinction is made, they record more infections than infected patients and more microbial causes of infection than infections. This is because a patient with one IaI, say of the urinary tract, is at increased risk of developing septicaemia and perhaps pneumonia as well if the causative microbe spreads into

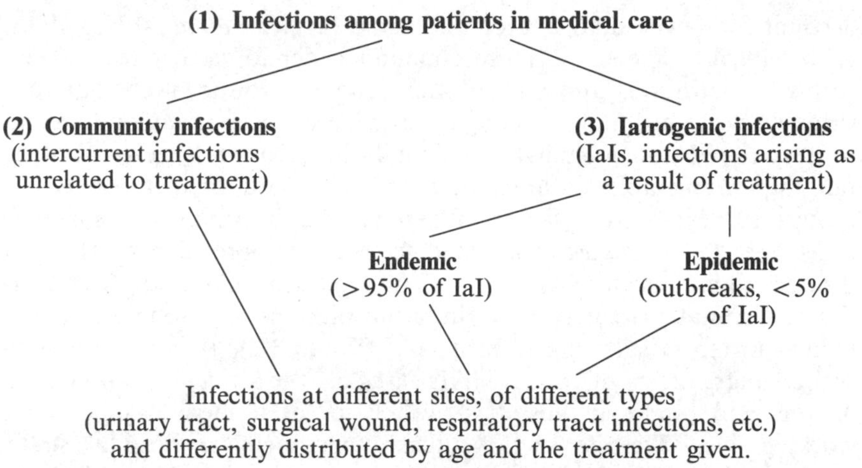

(1) Infections among patients in medical care

(2) Community infections
(intercurrent infections
unrelated to treatment)

(3) Iatrogenic infections
(IaIs, infections arising as
a result of treatment)

Endemic
(>95% of IaI)

Epidemic
(outbreaks, <5%
of IaI)

Infections at different sites, of different types
(urinary tract, surgical wound, respiratory tract infections, etc.)
and differently distributed by age and the treatment given.

Figure 3.2   The epidemiological varieties of infections that may be discovered among patients in medical care. The equivalent older terminology is, broadly, for (1) all infections in hospitals, for (2) community-acquired infections (CAIs), and for (3), hospital-acquired infections (HAIs). See p.26 for details of the differences between the terminologies.

the blood (Box 3.2). In this way, one patient may suffer from two or three different IaIs and these are counted as separate episodes. The size of the difference is significant, as can be judged from the figures given in Table 3.3. Similarly, a single infection is sometimes caused by two or more microbes acting together. For example, pus from a surgical wound may contain aerobic and anaerobic bacteria, both contributing to the pathology, so one case of HAI can contribute more than one microbe to the total of microbial causes.

THE DISTRIBUTION OF INFECTIONS

Iatrogenic infections present in two varieties: endemic (believed to be not less than 95% of all the infections that originate in hospitals) and epidemic (outbreaks, less than 5%). The infections that make up the endemic variety are distributed unevenly, though according to rules that can be deduced from a consideration of the different kinds of patients admitted to each unit or department and what is done to them there. The distribution of epidemic (outbreak) IaI is less predictable.

Endemic and epidemic infections

Hospital acquired (iatrogenic) infections (HAIs or IaIs) are divided into **endemic** and **epidemic** varieties (Figure 3.2). Although precise numerical data are scanty, epidemics (outbreaks) of HAI appear to account for fewer than 5% of the whole (Wenzel *et al.*, 1983; Haley *et al.*, 1985b). The much more common endemic variety (over 95%) forms a continuous and all-too-often ignored ground swell that runs through every hospital. Although individually these infections present in a scattered and seemingly random fashion, on a larger scale they fall into predictable patterns that become more predictable as the groups surveyed grow larger. When several hospitals are surveyed together, the averages that emerge for the prevalence of HAI produce figures such as those given in Tables 3.1 and 3.2. A rule-of-thumb estimate predicts that, in a hospital at any one time, one patient in ten is suffering from a HAI (Table 3.1). It is important to realize that most of these HAIs are of the *endemic* variety, so doctors and nurses in hospitals meet cases of it most days of their working lives. Despite such familiarity members of hospital staffs, most of whom should know better, fail to recognize these infections as iatrogenic and so lose sight of the fact that they are, to some extent, preventable.

The (largely endemic) infections described earlier in this chapter are not distributed evenly through individual hospitals. Patients in some units or departments routinely suffer from infections more often than those in others and the different types of infection are also spread unevenly. This unevenness is illustrated using information collected over a four-year period in the National University Hospital (NUH), Singapore (Meers and Leong, 1990). Hospital-acquired (iatrogenic) infections were surveyed by means of a partial incidence study. The partial nature of the study was imposed by a lack of the resources necessary to undertake a full incidence survey (p.89). Although the whole hospital was included, many cases that would only have been revealed by time-consuming reviews of individual patient records were not detected. A crude idea of the size of the shortfall is indicated by the difference between the overall rate detected in Singapore (3.4%) compared with the SENIC figure (5.7%) (Table 3.3). Despite the shortfall, the distributions of the types of infections found in Singapore are consistent with the other distributions in this table and in Table 3.2, so they may be used for illustrative purposes.

Where the endemic infections are found

Figure 3.3 shows where HAI was found in the NUH and the distribution of its types, by speciality. The greatest weight of infection fell on four of the six intensive care units. Patients in the worst affected areas suffered from HAI more than 16 times as often as those in the least affected ones, who occupied the beds allocated to the ophthalmic, ear nose and throat and dental surgery departments. Because the survey was only partial, absolute numbers of infections are not given in Figure 3.3, but the incidences shown, indicated by the heights of the columns, are proportional to each other. As noted, the general

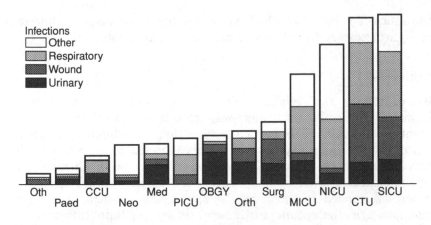

Figure 3.3 The amounts and types of hospital-acquired (iatrogenic) infections found in different units and departments in the National University Hospital, Singapore, in the four years 1986–1989 (Meers and Leong, 1990). The heights of the columns are directly proportional to the incidences of infection in each of them.

Key: Oth, other departments; Paed, paediatric department; CCU, coronary care unit; Neo, neonatal unit; Med, general medicine; PICU, paediatric intensive care unit; OBGY, obstetrics and gynaecology; Orth, orthopaedic department; Surg, general surgery; MICU, medical intensive care unit; NICU, neonatal intensive care unit; CTU, cardiothoracic unit; SICU, surgical intensive care unit.

distribution found agrees with that recorded in other advanced countries.

The special features of the distribution of infections and their variations by numbers and types will be discussed later. It is evident, however, that there is a comparatively heavy concentration of urinary tact infections (UTIs) among obstetric and gynaecological patients. This is a consequence of the anatomical location of the procedures carried out in this department and the extensive use of catheters. Catheters also accounted for the prominence of UTI in orthopaedic wards, where they were used in older patients confined to bed. There they equalled the numbers of surgical wound infections (SWIs), many of them associated with road accidents. Patients in general surgical beds suffered from UTIs less often, but lower respiratory infections (LRIs) appeared among them with greater frequency. This was a result of the effect of general anaesthesia (p.150) complicated in some cases by reduced respiratory movement after abdominal operations. The concentration of LRIs in intensive care units was due to the number of patients who required mechanical ventilation. Urinary tract infections were uncommon in the two paediatric intensive care units. This may be because no-one worries too much if a infant or child has a wet nappy or diaper, while it is thought necessary to catheterize an incontinent adult.

Outbreaks

By definition, two related infections constitute an outbreak. In hospitals outbreaks are commonly recognized as clusters of any number of cases of a particular infection (or different infections caused by the same microbe) that are connected in such a way that they appear to be related. The relationship may be in time, in space (their location), by the personnel who cared for them, by the procedure applied to them, by the application to them of a certain diagnostic or therapeutic instrument or by the application of the same batch of the same medication, by the observation that all the microbes responsible are identified as 'the same' or by a combination of these factors. Individually, outbreaks are unpredictable but, as with endemic HAI, when large groups of, say, paediatric patients are gathered together, sooner or later there will be outbreaks of viral gastroenteritis or of respiratory tract infection. Most doctors and nurses quickly recognize outbreaks of IaI for what they are. Although they may ignore the importance of the much more common endemic form of IaI, they tend to overreact to it when it appears in its epidemic form.

THE MICROBIAL CAUSES OF INFECTIONS

The microbes found in iatrogenic infections associated with an admission to hospital are described. Bacteria are the most common causes, but viruses, fungi, protozoa and even some worms may be involved. Non-bacterial microbes are of increasing importance among the growing population of patients whose immune systems are seriously disordered.

Algae and fungi in iatrogenic infections

The microbes least likely to cause any kind of infection are the algae. Although they produce toxins, algae, like plants, normally need light to grow, so cannot easily invade tissues or cause trouble under clothing or dressings. Fungi cause infections of two kinds. Those found in superficial infections, like ringworm or athlete's foot, can only grow in the horny layer of the skin or in the hair and nails. They rarely invade deeper tissues. Other fungi, most commonly *Candida albicans*, sometimes cause superficial infections in apparently healthy people. In those whose immunity is severely impaired, however, they can spread into the subcutaneous tissues to produce deep, spreading infections. The number of patients in this category grows year by year as more enterprising techniques are used to treat malignant disease, repair or replace damaged organs or as a result of the spreading pandemic of human immunodeficiency virus (HIV) infection and of its end stage, the acquired immune deficiency syndrome (AIDS). In these situations some other of the usually non-pathogenic fungi, such as *Aspergillus* spp., the agents of mucormycosis and *Pneumocystis carinii* (if indeed it is a fungus – see Edman *et al.*, 1988), may also cause serious infections.

Viruses in iatrogenic infections

Some respiratory viruses and enteroviruses cause sharp outbreaks of hospital-derived infections among niche groups of susceptible patients. Viral respiratory tract infections and gastroenteritis can be troublesome among infants and influenza is a significant cause of morbidity and mortality in geriatric units and nursing homes. Because viruses do not pose a special threat to the great mass of acute hospital patients, however, they are not conspicuous causes of hospital-acquired infections.

As with the fungi, this generalization fails among patients with impaired immunity who are more likely to be attacked, for example, by members of the herpes group of viruses (herpes, chickenpox, Epstein–Barr and the cytomegalovirus). Like other viruses, these cause primary infections in healthy people but unlike most others, they are not eliminated afterwards. They establish themselves permanently in a proportion of the cells they infect and hide there without (initially) doing any harm. This is called **latency**, a property that is all too evident to sufferers from repeated cold sores due to the herpes simplex virus. A patient with shingles experiences a recurrent infection with another member of the herpes group, the varicella-zoster (chickenpox) virus. This was left behind in latent form after an attack of chickenpox, often in childhood. Reactivation of the virus to cause shingles is more frequent and severe as the efficiency of the immune system is reduced. Debilitated elderly people, among them sufferers from malignant disease, are often nursed together. Shingles is much more likely to appear among such patients. If, by chance, two or three cases coexist, the impression is given of an outbreak. It has to be stressed that chickenpox (varicella) and shingles (herpes zoster) are part of the same infection, typically extended over many years. Either is infectious, but only to someone who has not previously had chickenpox (or to the very rare individual who fails to develop full immunity after a first attack of the disease). A non-immune person exposed to a case of chickenpox or of shingles may develop chickenpox. Exposure to shingles does not cause shingles.

Bacteria in iatrogenic infections

Bacteria are outstandingly the most common causes of infections that arise in hospitals. This is because large numbers of bacteria are already established on the body surfaces of both patients and members of staff (p.17). They are well placed to take advantage of the least, most localised immune defect. Damage to some of the components of the immune system is an almost inescapable consequence of admission to hospital. The damage may be local or general; for example, a local break in the surface of the skin or a mucous membrane or a general reduction in the number of white blood cells.

It is convenient to divide the bacteria that cause the more common varieties of IaI into the 'big three' major pathogens, and a miscellaneous group of 'others'. The big three are *Streptococcus pyogenes, Staphylococcus aureus* (Shanson, 1981; Goldman, 1992) and a number of Gram-negative rod-shaped bacteria (GNRs), whose similarities in the context of IaI allow them to be considered together as a single group. They include *Escherichia coli, Klebsiella* spp., *Pseudomonas aeruginosa, Enterobacter, Serratia, Citrobacter* and *Acinetobacter* spp., plus a few others. The characteristics that make the 'big three' important are summarized in Table 3.4.

At any one time medical and nursing staff in hospitals tend to perceive one of the 'big three' pathogens as a greater threat to patients than the others. For the time being this bacterium then attracts special attention and it achieves the notoriety of 'dominance'. In fact, it seems that GNRs have always caused more infections than the other two put together and staphylococcal infections have always been more common than streptococcal, even when the latter was regarded as the 'dominant' cause of HAI. The explanation of the apparent inconsistency is that, historically, the

Table 3.4 Some of the characteristics that contribute to the importance of the major bacterial pathogens in iatrogenic infections

	Strep. pyogenes	*Staph. aureus*	*Gram-negative rods*
Normal habitat	Pharynx	Nose, moist skin	Gut
Colonization frequency	< 5–10%	20–50%	100%
Resistance to drying*	Fair	Good	Most, poor
Virulence	High	Moderate	Moderate
Resistance to antimicrobials	Unimportant	Very important	Important
Typical sites of IaIs	Skin, wounds, septicaemia	Skin, wounds, septicaemia	Urinary and respiratory tracts, wounds, septicaemia

* See text.

streptococcus has been most feared. Although not identified until the science of bacteriology was founded in the latter half of the 19th century, clinical descriptions of infections clearly indicate that it was active earlier. When prevalent, the streptococcus is always dominant because of its extreme virulence. It can kill apparently healthy people in a few hours and still occasionally does so. Other members of the 'big three' do not do this. The introduction of the sulphona-mides in the 1930s accelerated what already may have been the beginning of the decline of the streptococcus and it ceased to be important as a cause of HAI in the 1940s, after the appearance of penicillin. It would be rash to assume that the streptococcus will not one day return as a common cause of infections, including those that arise in hospitals. If this were accompanied by the acquisition of resistance to penicillin, the result would be a problem that would once more dwarf any concern about staphylococci.

Streptococci in iatrogenic infections

In the 1930s and 1940s, when streptococci were still the dominant pathogens, work on the epidemiology of hospital infections led to the imposition of a number of infection control measures (Chapter 2). One was the practice of screening people bacteriologically for the presence of *Strep. pyogenes*. Members of staff who were swabbed and found to be carriers were excluded from work. Carrier patients were isolated. This was possible due to the low rate of carriage of this microbe, so the number of individuals involved at any one time was not large and the disruption was acceptable. Many more practices were founded on the discovery that the inanimate environ-ment surrounding patients suffering from streptococcal infections was plentifully contaminated with their streptococci. This gave rise to the idea that bacteriological surveillance of the environment was a useful way of detecting contamination that was thought to be the cause of infection. It also led to the introduction and enforcement of such things as damp dusting, the sterilization of bedding, the widespread use of antiseptics and disinfectants, the 'no-touch' technique for dressing wounds, the use of filtered air in operating rooms and elsewhere and the wearing of masks outside operating departments.

None of these measures was introduced in a controlled fashion, so their effectiveness could not be measured. The search for carriers has some logic but, because the necessary bacteriological examination takes not less than 24 hours to complete, carriers are likely to have passed on their microbes to others before they are detected and removed from circulation. The discovery that streptococci exposed to the air rapidly lose their infectivity (p.50) throws doubt on the usefulness of the environmental measures. Unfortunately once they had been introduced they rapidly acquired the force of essential rituals. Without critical examination, some of them are still used to protect patients from a risk of unknown size, due to microbes for which they were not designed and against which they are not known to be effective.

Staphylococci in iatrogenic infections

The growing availability of penicillin in the late 1940s was accompanied by the increased prevalence of staphylococci that were resistant to it due to the production of the enzyme beta-lactamase (penicillinase). These 'hospital' staphylococci rapidly replaced streptococci as the dominant causes of HAI. By 1948 they accounted for 60% of the staphylococci that were isolated in some hospitals in the UK; by 1955 figures as high as 90% were reported. Outbreaks of infection due to it were detected all over the world.

In 1960 methicillin, a form of penicillin resistant to beta-lactamase, was introduced. This was active against the hospital staphylococcus and the 'pandemic' disappeared (pp.13,14). Almost immediately the first methicillin-resistant strains of *Staph. aureus* (MRSA) were identified. Perhaps because methicillin and its successors (cloxacillin and flucloxacillin) were used more rationally and in smaller quantities than had been the case with penicillin, MRSA spread much more slowly than had been the case with the hospital staphylococcus in the 1940s and 1950s. Initially it did not achieve much prominence, but this changed in 1976 when strains of MRSA that were also resistant to the aminoglycoside antimicrobial gentamicin (MARSA) appeared and began to spread. At the time of writing, MRSA is distributed internationally, in varying abundance. It has not yet duplicated the highest efficiency of the 1950s pandemic strain to replace less resistant staphylococci, nor has it acquired its alleged superior virulence (though see p.165). Available evidence suggests that when ordinarily resistant *Staph. aureus* (ORSA) and MRSA coexist, the total number of cases of staphylococcal infection does not rise, but that infections with MRSA displace some of those due to ORSA, within the same total. This is consistent with the view that ORSA and MRSA are not biologically distinguishable as commensals and potential pathogens, other than as determined by the application of the artificial selective pressure of antimicrobial drugs. MRSA causes concern because it is necessary to use more expensive and perhaps toxic antimicrobial drugs in the treatment of infections due to it. It attracts more attention in some places than in others and when it does so, it causes expensive disruption disproportionate to its real importance.

In Singapore MRSA established itself at the same time as the third-generation cephalosporins were introduced into clinical practice as first-line therapeutic agents. MRSA quickly achieved a 50% share of all the infections caused by staphylococci. Although anecdotal, there is a strong similarity between this evolution and that of the hospital staphylococcus in the early 1950s. In both situations staphylococci were exposed to similar selection processes, with the same outcome. It is notable that Singapore is one of the many countries where MRSA has been largely ignored where, although it is widespread, it had little or no effect on hospital practice (Meers and Leong, 1990).

Gram-negative rods in iatrogenic infections

In contrast, GNRs probably attract less attention than they deserve. Although almost certainly the most common causes of IaI, it was only between 1960 and 1976 that GNRs finally achieved the notoriety of dominance. They did this when streptococcal infections had disappeared and staphylococci were no longer considered a threat. Multiply resistant strains of different GNRs were observed to cause limited outbreaks of IaI in some hospitals (see, for example, Casewell and Phillips, 1978; Curie *et al.*, 1978; Chow *et al.*, 1979; Pitt, Erdman and Bucher, 1980; Cross *et al.*, 1983). When the staphylococcus re-emerged in the form of MARSA, however, GNRs fell back into comparative obscurity, though they had not gone away. Figure 3.4. and Table 3.5 show that most cases of IaI are still

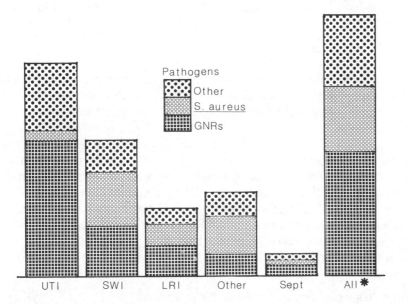

Figure 3.4 The distribution of the major bacterial pathogens causing different hospital-acquired (iatrogenic) infections in the National University Hospital, Singapore, from January 1986 to December 1989.

Key: UTI, urinary tract infection; SWI, surgical wound infection; LRI, lower respiratory infection, Other, other infections; Sept, septicaemia; All, all infections, drawn to a scale 50% of the others; GNR, Gram-negative rod-shaped bacteria.

Table 3.5 The distribution of the major bacterial pathogens causing hospital-acquired (iatrogenic) infections in the National University Hospital, Singapore, from January 1986 to December 1989, as they were distributed by speciality

	HAI - causative organisms, numbers (%)				
	GNR	ORSA	MRSA	Others	Totals
Surgical intensive care	99 (52)	8 (4)	30 (16)	53 (28)	190
Cardiothoracic unit	46 (48)	9 (9)	18 (19)	22 (23)	95
Neonatal intensive care	39 (20)	29 (15)	66 (33)	65 (33)	199
Medical intensive care	67 (42)	12 (8)	26 (16)	54 (34)	159
General surgery	503 (51)	97 (10)	130 (13)	267 (27)	997
Orthopaedics	236 (52)	39 (9)	90 (20)	85 (19)	450
Obstetrics & gynaecology	391 (46)	122 (14)	7 (1)	323 (38)	843
Paediatric intensive care	8 (28)	4 (14)	9 (31)	8 (28)	29
General medicine	463 (57)	62 (8)	102 (12)	192 (23)	819
Neonatal unit	31 (21)	66 (45)	16 (11)	33 (23)	146
Coronary care	17 (53)	3 (9)	3 (9)	9 (28)	32
Paediatrics	63 (40)	32 (20)	22 (14)	40 (25)	157
Others*	11 (35)	8 (26)	3 (10)	9 (29)	31
Totals	1974 (48)	491 (12)	522 (13)	1160 (28)	4147

Key: *Ophthalmology, ear, nose and throat & oral surgery; GNR, Gram-negative rod; ORSA, methicillin-sensitive, ordinarily-resistant *Staph. aureus*; MRSA, methicillin-resistant *Staph. aureus*.

caused by GNRs. Proportionately they cause septicaemia more often than staphylococci. If the ability of a microbe to cause septicaemia is a measure of its pathogenicity, GNRs are no less virulent than staphylococci.

Figure 3.4 and Table 3.5 illustrate the distribution of the major bacterial pathogens by types of infection and speciality, respectively. Because of current interest, distinction is made in the table between methicillin-sensitive *Staph. aureus* (ORSA) and MRSA. The data are from the NUH, Singapore (p.32). With the exception of the high incidence of infections due to MARSA, the relative frequencies of the different pathogens and their distributions agree with similar statistics from other countries.

Other 'bacterial' causes of iatrogenic infections

The principal members of the miscellaneous group of bacterial causes of IaIs ('others' in Table 3.5) are streptococci of Group D (mainly members of the new genus *Enterococcus*), *Candida* spp., *Staph. epidermidis* and a variety of anaerobes. These were found as common causes of urinary tract infections and numbers of them were isolated from infected wounds and infections of the respiratory tract. *Staph. epidermidis* has been found on its own as a genuine pathogen in infections associated with central vascular lines, urinary catheters and the catheters used for peritoneal dialysis. It is not infrequently reported in cases of septicaemia. Like *Staph. aureus*, it has developed resistance to methicillin. Bacteria that have emerged in recent years in hospitals as notable pathogens are *Clostridium difficile*, vancomycin-resistant enterococci and in some places multiply resistant *Mycobacterium tuberculosis* (p.164).

'Hospital strains'

With the partial exception of *Strep. pyogenes*, all the bacteria that have achieved dominance or have otherwise become prominent as causes of IaIs have been naturally resistant or have acquired resistance to one or more important antimicrobial drugs. This is the principal characteristic that has led to their recognition as 'hospital strains' of bacteria and in some places to their listing as one of the 'alert organisms', the appearance of any of which triggers a response by the infection control organization concerned (p.91).

The typing of bacteria

Because infections due to the major pathogens are so common, if, say, three cases of IaI have been caused by *Esch. coli* it does not follow that the infections are related to each other. The identification of an outbreak of infection due to such a common pathogen depends on reasonable proof that the three strains of *Esch. coli* are, as nearly as it is possible to determine, 'the same'. To illustrate this, suppose that, over the course of a few weeks, there have been five cases of staphylococcal wound infections following operations performed by the same surgical team. On investigation, two members of the team are found to be staphylococcal carriers. Which, if either, might be responsible? As up to 50% of normal people may carry *Staph. aureus* at some time or other, both may be innocent because unless the cases and their bacterial causes are connected, there is no outbreak. It is necessary to examine the staphylococci from the five patients and the two members of staff by methods designed to reveal minor differences. If they are all different there is no outbreak. If two or more of them are indistinguishable, the necessary inferences are drawn and precise action can follow. To answer questions like this various methods have been developed that can reliably determine if two or more strains of most of the important causes of IaI are distinguishable or indistinguishable. Unfortunately these 'typing schemes' nearly always require scarce or expensive reagents or skills that confine them to reference laboratories. The need to refer specimens to distant laboratories means that urgent decisions can be delayed.

To overcome this, microbiology laboratories in hospitals make use of the results of their biological identification tests and antimicrobial sensitivities. These often allow a preliminary judgement to be made on relatedness so that a decision can be taken whether to proceed with an investigation, take rapid remedial action or reject the idea of an outbreak. If two or more strains give different results in these routine tests they are unlikely to be related and there is little possibility that they come from patients involved in an outbreak of infection. If they are indistinguishable it is sensible to accept that an outbreak might exist. Strains should of course be kept for submission to a reference laboratory for final adjudication, if the circumstances require it.

Iatrogenic infections and the immunodepressed

Different patterns of infection arise among the comparatively small number of patients who are seriously immunodepressed and the microbiology of these infections is also different. When admitted to hospitals such patients tend to congregate in transplant and oncology units or in places where patients with AIDS are cared for. With increasingly severe immunological deficiency less pathogenic microbes emerge as serious pathogens. Among the herpes viruses the less pathogenic cytomegalo- and Epstein–Barr viruses may re-emerge to cause serious disease. This pattern is repeated with the fungi, a few of the protozoa and even some of the worms that infest man. *Pneumocystis carinii* and the protozoon *Toxoplasma gondii* are examples. Many people share the latter parasite with their cats, nearly always without knowing it. This is another latent infection that may be activated and cause serious disease in patients whose immunity is impaired.

Patients who receive transplanted organs or other tissues are immunosuppressed to prevent the rejection of their transplants. Latent microbes may already be present in the recipient, or, more dangerously, they may be transferred in the donated tissue or a blood transfusion to a recipient who has not previously encountered the microbe in question. In either case the immunosuppression necessary to preserve the transplant may also allow the latent microbe to reactivate, with potentially serious consequences.

An increasingly common cause of immunodeficiency is the human immunodeficiency virus (HIV). Other than historically in the case of blood transfusions and haemophiliacs who received factor eight derived from blood from carriers of the virus, the infection is nearly always acquired in the community, but its end stage, AIDS, often presents in hospitals. AIDS develops after a period of months or years during which the HIV has attacked and destroyed important parts of patients' immune systems. They are then subject to repeated attacks of infections due to one or more of the microbes already mentioned, plus others that take advantage of their immunodepression. They may also develop forms of cancer that are normally suppressed by the immune system. Most of the infections suffered by patients with AIDS are caused by microbes that are weakly or completely non-pathogenic for other people. They are not a threat to healthy members of hospital staff. Many healthcare workers have been concerned that they might acquire an HIV infection occupationally. Extensive experience has shown that this is very rare and that it is largely avoidable (p.236). Tuberculosis is an important exception to this rule. There is a genuine risk of serious occupational infections among staff exposed to patients with this disease, particularly when resistant strains of the bacterium are involved (pp.172 and 236).

THE TRANSMISSION OF INFECTIONS

The terms source, reservoir, point of exit, pathway, vector, portal of entry, endogenous, exogenous, self, cross and environmental are defined, as applied to the epidemiology of iatrogenic infections. How, when and why the factors associated with these terms contribute to the genesis of infection.

Definitions

To cause an infection a microbe must have come from somewhere, a **source of infection**. Sometimes it has to find a **point of exit** to escape from the source and in every case the microbe requires a **vector** or **vehicle** of infection to conduct it in sufficient numbers (an **infectious dose**) along a **pathway** to reach a **portal of entry** on or in the new host. The portal of entry is the location at which the microbe attaches to the host, to establish a foothold or bridgehead from which to develop into an infection. In natural infections the portal of entry may be the respiratory or alimentary tracts, skin or a mucous membrane or the placenta in an infection of a fetus. Diagnostic and therapeutic procedures provide a number of highly efficient additional, artificial portals of entry for the direct introduction of microbes into the body, to cause iatrogenic infections.

A **reservoir** of infection is a place in which an infecting microbe is found. It may also be the source of an infection but only if there is a point of exit from which it can escape and there is an effective vector to lead it to a potential host. The waste traps of sinks, for example, usually contain large numbers of microbes. In clinical areas these will include pathogenic bacteria washed from the hands of staff members. Sink waste traps are reservoirs of pathogenic microbes, but it is difficult to imagine circumstances in which they might escape and be transferred to be the source of an infection. This example can be multiplied many times. Time and money are saved if reservoirs of microbes (like sink waste traps) are ignored unless they are also credible sources (Box 3.5).

When looking for the origin of an infection, it is logical to search

---

Box 3.5   Microbes = infection?

Most members of the public would accept the direct, oversimplified relationship suggested in the heading to this box. A glance through a few back numbers of, for example, the *Journal of Hospital Infection* shows that it is not uncommon for modern investigators in the field of infection control to do the same thing. Some justify their assumption of a direct link by reference to prior publications, perhaps sanctified by age, which may describe experiments the details of which they do not re-examine for the possibility that the original data were misinterpreted.

   Control measures proposed in papers whose authors do not set out to *prove* the existence of a direct *causal* relationship between the (usually environmental) microbes they describe and *real* infections should be assessed with more than a single grain of sceptical salt!

---

for the nearest source where microbes indistinguishable from those responsible are present in sufficient numbers to provide an infectious dose (p.21). If more than one potential source is located, it is wise to study first that most abundantly supplied with the microbes concerned. Reservoirs should be recognized for what they are and avoided. Although biology has a way of ignoring oversimplifications the application of this rule will, in many cases, produce a working hypothesis and may lead to the right answer as well.

Sources

It is customary and useful to divide all the possible sources of infecting microbes into two classes and three categories (Figure 3.5). Following the rule just set out, in the case of a patient infected with a common organism, the nearest point at which they are most abundant

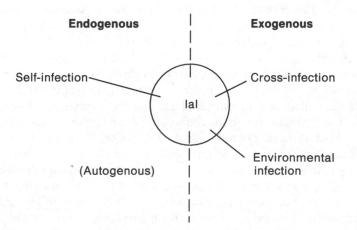

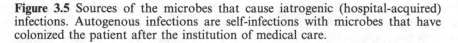

Figure 3.5 Sources of the microbes that cause iatrogenic (hospital-acquired) infections. Autogenous infections are self-infections with microbes that have colonized the patient after the institution of medical care.

is likely to be the surface of their own body. Infections that originate from this source are **endogenous** or **self-infections**. It was pointed out earlier in this chapter that, after admission to hospital, many patients add hospital strains of bacteria to their normal floras. If a self-infection then develops and the pathogen concerned has been added in this way, the infection may be called **autogenous** rather than endogenous.

The next most likely sources of infecting microbes are the surfaces of the bodies of other patients or members of the healthcare workers. If one of these is the source the result is both exogenous and, because it originates in someone rather than something else, a **cross-infection**. The third possibility is that the infecting microbes come from some part of the inanimate environment. These are also exogenous, but in this case are **environmental infections**. Because the measures appropriate to the control of IaIs differ according to their source, it is important to distinguish between self, cross and environmental infections. Self-infections in particular cannot be prevented by measures appropriate to environmental infections. It is a waste of time and money to apply them.

Self-infections

Self-infections are due to microbes already present on or in the host before the new disease process begins. In the case of bacterial infections the source may be one of the normally colonized surfaces of the body, an abnormally colonized privileged surface (p.18) or a coexistent infection at another site. The latter is a particularly significant and efficient source. It ensures that a patient who already suffers from one form of iatrogenic infection is more likely to develop a second or a third (p.29). The vector (pathway) in this case may be by direct spread as microbes extend over or through the patient's own tissues, bloodstream or lymphatics or by transfer from one part of the body to another on the hands of the patient or an attendant.

The normally colonized surfaces of the body are made up of the skin, the respiratory tract above the larynx, the gastrointestinal tract from mouth to anus and, in females, the vagina (p.17). The types and numbers of bacteria that occupy these sites vary, but they are always numerous. The **skin** is the largest organ of the body. It contains hair follicles, sweat and sebaceous glands and it includes the hair and the nails. It varies in thickness, the extent to which the surface is cornified and the degree to which it is moist (groin, perineum, axillae and flexures) or more dry (exposed areas). The surface is constantly shed at a rate that varies from place to place on the body and from person to person. In general men shed more than women. Shedding is of flattened leaf-like squames, about 10–20 μ across, which may be detached individually or in groups. When large numbers adhere together desquamation is a visible process, as with dandruff or scaly skin. Significantly more squames are shed immediately after washing and drying, particularly if more powerful detergents are used in place of soap. Most squames are sterile, but a variable proportion act as rafts for one or more living bacteria. An average adult loses about 300 million squames a day. If one in a hundred of them carries bacteria, an individual may shed 2000 contaminated particles every minute. In enclosed occupied spaces nearly all the bacteria in the air and much of the dust originate in this way. Minute fibres rubbed from clothing make up a large part of the remainder. Nearly all are sterile (Selwyn and Ellis, 1972; Meers and Yeo, 1978; Noble, 1983).

The **normal resident flora** of the skin is composed largely of aerobic Gram-positive staphylococci and corynebacteria, though anaerobes are numerous in some areas and Gram-negative bacteria are found in the moister parts. Their total numbers vary on different parts of the body between 4000 and 400 000 in each square centimetre (Noble, 1983). *Staphylococcus aureus* does not often reside on dry, healthy skin but it is regularly found on the moist parts of up to 50% of normal individuals, in whom it is often simultaneously present in the nose. When found on the skin it favours a site in hair follicles just below the surface, where it may be missed by ordinary swabbing for bacteriological purposes, and can survive the application of an antiseptic.

Resistance to (or exclusion of) new bacterial colonists (p.22) is a particular feature of normal skin. This is achieved in a slightly acid environment with the active co-operation of the resident, normal or permanent flora. Parts of the body regularly come into contact with more or less heavily contaminated surfaces. When foreign bacteria are transferred to the skin by such contacts, some time (perhaps hours) must elapse before the defences dispose of the temporary residents. In the interval the foreigners form a **transient flora**. Faeces is one of the most heavily contaminated materials to act as the donor of a transient flora, followed by the surfaces of one's own or other people's bodies. Ordinary contacts between people, particularly in hospitals, most often involve the hands, so these acquire new transient floras perhaps hundreds of times a day. While these transient microbes survive, they may be transferred to another individual, again by contact. This is why hands are important as vectors in cross-infections (see below).

Most of the transient flora is readily removed by ordinary washing. This includes any *Staph. aureus* that individuals have transferred from their own noses to their hands. The importance of such transfers can be determined by watching how often people touch their faces without realizing it. By contrast, the resident flora is much more tenacious, because the microbes concerned spread through the thickness of the cornified layer and down into glands and hair follicles. Washing reduces their number, particularly if antiseptics are used, but the resident flora cannot be removed altogether by any treatment that leaves skin alive. If the skin is damaged or diseased bacteria that might have been transient may now find conditions suitable for colonization or infection and so become part of the resident flora. Exposed eczematous skin may harbour a million colonizing *Staph. aureus* per square centimetre and if the eczema is infected the number may rise tenfold (Leyden, Marples and Kligman, 1974). Healthcare workers in this condition are a danger to patients.

Hair may be a crowning glory, but it has a curiously bad microbiological image. The scale of preoperative shaving that was and in some places still is thought necessary is an indication that hair is considered to be an infection hazard. In fact, hair is less able to support microbial multiplication than is the skin from which it grows. Of course, dense hair keeps the underlying skin moist and so encourages microbial growth. In these circumstances damp, heavily colonized skin squames separate and are trapped among the hair over it. To this extent hair is 'dirty', but this does not apply to areas with sparse or short hair or to hair that is clean. Part of the bad image may be due to the fact that hair and skin squames are both shed, but the result with hair is more obvious and so objectionable. The finding of a hair in the soup is a classic culinary disaster. Fortunately we are not so easily reminded that every bowl of soup (and everything else we eat) must contain large numbers of invisible squames derived from the skin of the cook who prepared it.

The **upper respiratory tract**, particularly the mouth and throat, provides a damper, warmer environment in which streptococci are more at home. Here they are joined by increasing numbers of anaerobic and Gram-negative bacteria. It is popularly supposed that these bacteria escape into the air in large numbers during breathing and speaking and of course it is well known that 'coughs and sneezes spread diseases'. In fact, several experiments have shown that many respiratory infections are spread more effectively by such physical contacts as holding a baby or shaking hands (Wenzel, Deal and Hendley, 1977; Hall, Douglas and Geiman, 1980; Couch, 1984). They are spread less effectively by handling items recently used by an infected person and rather poorly by sitting with them in the same room, with no physical contact. Coughs and sneezes may spread diseases, but the efficiency by which they do so is in doubt. The subject is clouded by strongly held beliefs about the importance of the air as a source of infection (Anon, 1988; Chapter 2).

The **gastrointestinal (GI) canal**, particularly its lower part, is where much of the normal flora of the body is found. This wet environment lacks oxygen so it favours anaerobes and several varieties of facultatively anaerobic Gram-negative rods. Among these, strains of *Escherichia coli* are the most common, so it is not surprising to find that these are also the most common causes of iatrogenic infections. As noted, the normal flora of the body may change following admission to hospital. How and why this is responsible for self-infections with hospital strains of bacteria is described earlier in this Chapter, together with the growing importance of self-infections due to various microbes capable of latency. Many of these are viruses, though fungi and protozoa may also play a part.

Cross-infections

The microbes responsible for cases of iatrogenic cross-infection are derived directly from the bodies of other people, either patients, staff or visitors. Because body surfaces are involved they have much in common with self-infections. and indeed, may be indistinguishable from them. Where they differ is that they are more likely to be due to hospital strains of bacteria. They may be indistinguishable because it is usually impossible to say if a particular infecting organism was transferred to the site of the infection at the moment it was initiated or beforehand, so had become part of the patient's flora before the infection started. Because a colonization (however brief) always precedes an infection as its necessary incubation period, the distinction between an autogenous self-infection and a cross-infection is, in fact, arbitrary. It is not, however, unimportant. The existence of the autogenous route in the development of self-infections explains the mode of transmission in many cases of IaI. For the reasons noted above, it is believed that hands are the principal vectors in these cases as well as in cross-infections.

Environmental infections

The parts of the environment that may be directly associated with iatrogenic infections are shown in Figure 3.6. It is unusual for the inanimate environment to produce, on its own, microbes that are pathogenic for humans. In most cases the environmental contamination originated in some other living thing, most often people, animals or plants. **Legionellosis** (legionnaires' disease) appears to be an example of a wholly environmental infection, though in some cases at least, *Legionella pneumophila* in fact multiplies inside amoebae that inhabit the same ecological niche. Legionellosis does not pass from person to person, so a patient suffering from it need not be isolated. Legionellas grow in water and proliferate actively in certain human artefacts, principally air-conditioning cooling towers of the wet variety and complex plumbing systems. Large modern buildings are particularly vulnerable, so hotels and hospitals are quite often implicated in outbreaks of the disease. The ecological niche must provide the right temperature (neither too hot nor too cold), plenty of organic matter and a vehicle or vector of infection. In the case of cooling towers the vector is the fine mist or aerosol of water they spray into the air. Any legionellas this contains may be inhaled. Although cooling towers are sited on the outside of buildings, the spray from them may be drawn inside through an air intake or a

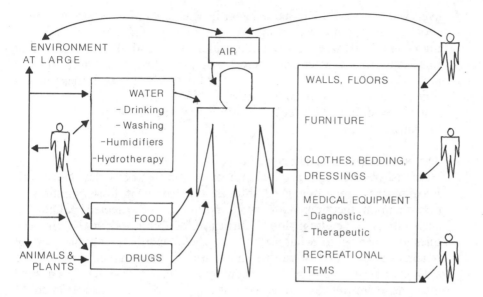

Figure 3.6 A summary of the more important potential sources of environmental infections.

window. Patients in hospital may be exposed to water from contaminated plumbing systems when showering or bathing or even from water in humidifiers or chairside dental apparatus.

Aspergillosis and mucormycosis are forms of IaI that are more incontrovertibly environmental. *Aspergillus* spores are ubiquitous and they have recently emerged as a cause of iatrogenic and community infections among the growing population of severely immunocompromised patients. Formerly they were recognized as little more than causes of allergy. Free-living organisms of the *Pseudomonas* and *Acinetobacter* groups grow in damp environmental sites and may spread from these to cause colonizations and perhaps IaIs. There is evidence that strains of these bacteria freshly isolated from the environment are less pathogenic than those that have recently caused infections and so are already adapted to the human body.

Animals (including birds and fishes) and plants are connected to patients as providers of food and less commonly through certain drugs, including the ingredients of some intravenous infusions. Microbes able to cause IaIs may pass along the same chain. Many of the standard microbial causes of food poisoning are of animal origin. They are introduced into hospitals in food and regularly cause IaI, sometimes with the inadvertent connivance of an inefficient kitchen. Animals and birds may also contaminate water supplies. In some tropical countries insects may be a real problem as vectors of infections like malaria. Such scavengers as semiwild cats, rodents, flies, cockroaches, ants and other winged and creeping arthropods are everywhere perceived to be a hazard. The reputation is largely unde-served, but it is an interesting example of the confusion that exists in many minds between aesthetic considerations and the real causes of infections. In a kitchen the presence of these pests is an indication of a low standard of hygiene or poor facilities or both. It is these inade-quacies that are the primary causes of food poisoning, not the pests. They are doing their best to clean up the mess made by dirty humans.

The environment as a source of infection

Humans are the most prolific sources of the microbes that cause IaI. Because they are often shed into the environment it is common to find them in the immediate surroundings of an infected patient on such things as bedding, floors, furniture, clothing and in the air. The question then is to what extent are these microbes capable first of reaching and second of causing infections in other patients?

One of four things may happen to a microbe that has been shed into the environment. Most pathogenic bacteria are accustomed to the five-star accommodation provided for them on or in the human body. When they leave they enter a cold hard world where they must find alternative sources of nutrition, most critically of water, otherwise most of them die more or less quickly. This is particularly true of the Gram-negative organisms that live on the most sheltered body surface, the gut. Bacteria from drier surfaces of the body withstand desiccation a little better, so survive longer when they are shed from it. The length of time these organisms take to die varies according to the temperature, humidity, the presence or absence of sunlight (equals ultraviolet light), and whether or not they are coated with protective organic matter. It also varies according to the energy put into their recovery. As with other living things, microbes rarely die instantaneously. If microbes are resuscitated in the laboratory equivalent of a human intensive care unit, bacteria that would otherwise be pronounced dead may recover. Whether such laboratory manipulation has any relevance to infection is not clear, but the time taken for most ordinary bacteria to die, as traditionally established, is measured in minutes to hours.

Microbes lose virulence before they die

The survival of a microbe is not the same as its ability to cause an infection. Fully virulent *Streptococcus pyogenes* that have been exposed to the air for some time can still grow in the laboratory, but they have lost their capacity to cause infections (p.12). A number of experiments have been done to prove this, in one of which up to a million living but partly dried streptococci were blown directly into the throats of volunteers (Rammelkamp, Mortimer and Wolinsky, 1964). None of them was infected. A control group of volunteers similarly exposed to an identical preparation that had not been dried developed typical streptococcal tonsillitis. *Staph. aureus* has been found to behave similarly (Hinton, Maltman and Orr, 1960) and some Gram-negative rods exposed to chlorhexidine, although still completely viable in the laboratory, were found to be of reduced virulence (Holloway, Bucknall and Denton, 1986). In both the latter experiments human volunteers were replaced by mice!

In the case of *Strep. pyogenes*, a mechanism is available to explain this observation. Fully virulent streptococci can repulse an attack by polymorphonuclear leucocytes (PML), which are an important part of the body's innate system of immunity. Each *Strep. pyogenes* has on its surface one of over 70 different kinds of delicate, frond-like proteins, collectively called the 'M' proteins. These defend streptococci from attack by PML. In the course of an infection an individual develops antibodies to the 'M' protein of the attacking streptococcus. The next time the individual meets the same 'M'-type streptococcus, the antibody combines with the 'M' protein and inactivates it, so that PML can now attack and destroy the invader. Drying damages the delicate fronds of the M protein, so the organism's defences against PML are breached.

In the early stages of an infection microbes must attach themselves to a surface on or in a host and then defend themselves against the host's immune systems. As with streptococci, the mechanisms that enable them to do this are located on their surfaces. Minor, non-lethal, superficial damage must therefore reduce a microbe's ability to initiate an infection. As noted above, a partially desiccated streptococcus is a cripple that can no longer infect a surface or tissue where PML are active. In the case of the other microbes mentioned it is probable that non-lethal superficial damage is also responsible for their reduced virulence.

Some bacteria and most fungi have particular properties that enable them to withstand adverse conditions for a long time. The best examples are the microbes that form seed-like spores. These survive and retain their pathogenicity for months or years without water or nutrients and can often withstand physical and chemical disinfectants as well. The causes of anthrax, tetanus, botulism, gas gangrene and pseudomembranous colitis (*Clostridium difficile*) all belong to spore-bearing families of bacteria. Bacteria that do not form spores but with similar properties are *Coxiella burnetii* (the cause of Q fever) and the tubercle bacillus. The tubercle bacillus is an important cause of iatrogenic infections. *Coxiella burnetii* is not.

Tuberculosis as an environmental infection
The pulmonary variety of tuberculosis is its most common form. It is acquired when tubercle bacilli (*Mycobacterium tuberculosis*) are inhaled deep into the lungs. To slip through the tortuous respiratory passages lined with sticky mucus, the particles containing the bacilli must be very small, about 5 μ in diameter or less. When coughed up by a person with open tuberculosis, most of the bacilli are enmeshed in large gobbets of more or less viscous mucus. It is difficult for

particles of the size that might cause infections to be formed at this stage. Comparatively large wet particles of sputum fall onto the floor or onto other surfaces where they dry and then break up into smaller fragments. If these are swept up into the air they are available for inhalation and may be of the right size to cause an infection. Most other microbes trapped in particles of sputum desiccate with the mucus, so are inactivated or die before they are able to do any harm. Tubercle bacilli are different. They are very resistant to drying, so survive the process. This is how this pathogen is delivered to exactly the right place to establish another tuberculous infection. Pulmonary tuberculosis is an environmental infection with airborne transmission as its principal mode of spread and man is the source of the contamination of the environment. The same may be true of infections caused by *C. difficile* (p.171). There are not many other examples.

Debunking the environment as a source of infection
Because it is a common perception that the environment is an important cause of infection and particularly of IaI, it is necessary to elaborate this point. Reference has been made (p.6) to Semmelweis' evidence that cross-infection is more important than environmental infection as a cause of IaI. Much more recently the transfer of a teaching hospital to a new building provided evidence in support of this. The infection rate among patients and the bacteriological conditions of the two environments were monitored continuously before and after the move. The initially clean environment in the new hospital deteriorated bacteriologically after occupation, taking 6–12 months to reach a state equivalent to the old one. Despite the cleaner environment enjoyed by patients in the first months of occupation of the new hospital, infection rates remained exactly the same (Maki *et al.*, 1982). Another experiment tending to the same conclusion will be described later in the section dealing with wound infections (p.139). The principal evidence leading to a contrary view arose from work done in burns units, where conditions are uniquely different (pp.13 and 224).

Perception and reality are at variance in this case because to establish an infection a finite and often quite large population of fully virulent microbes must find their way to a portal of entry on a patient. They must then establish a colonization before an infection can develop. Microbes in the environment find this difficult for the reasons just described, but their principal obstacle is a numerical one. Microbes cannot multiply in dry environments (and most of the hospital environment is dry) and once they have arrived in it most microbes lose their pathogenicity and die off more or less quickly. There is an enormous disparity in numbers of bacteria between the dry environment and body surfaces. If the bacteria carried by an average person (10^{14}–10^{15}) were laid one in each of the millimetres that separate the earth and the sun, the line of bacteria would make the trip all the way there at least once. By contrast the number of bacteria contained in the whole volume of air in a very unsatisfactory operating room while in use similarly laid out would reach the top of a very small hill (about 60 m). The number of bacteria on the whole floor of the same OR might stretch 5 km and a floor of similar size that would have been judged hygienically poor when environmental testing was in vogue might provide enough to stretch 40 km. Confronted by such numerical differences the logic is inescapable. Self- and cross-infections are far more important causes of IaI than anything that might originate in the dry environment. It is not surprising that applying disinfectants to floors or walls or spraying them into the air has never been proved to be of any use.

Further support for the view that body surfaces are the most important sources of IaIs arises from the observation that the frequencies of their main bacterial causes agree closely with the frequencies with which they are found as contributors to the body's normal flora (Tables 3.4 and 3.5; pp.36,40). *Esch. coli* (plus other Gram-negative enterobacteria) are much the most common causes of iatrogenic infections and they are also carried by 100% of the population. *Staph. aureus*, carried by up to 50% of the population, is the next most common pathogen. In modern conditions outside burns units *Strep. pyogenes* is carried much less commonly, so it is a rare cause of iatrogenic infections.

The situation is totally different for those parts of the environment that are wet. Food and drink are potentially important sources of the bacteria and viruses that may cause IaI. The same applies to drugs used in liquid form. Intravenous infusions, eye drops and disinfecting solutions at 'in use' concentrations all have bad reputations, as do pieces of diagnostic and therapeutic equipment that contain liquid or that may collect it in the form of condensation. These are all potential sources of infection as highly efficient vectors or vehicles. These and other examples of environmental hazards associated with wet environments are considered in Chapter 10, p.229. The precautions quite properly applied to prevent harm in these cases should not be extended to other situations. Once more it is necessary to make a distinction between sources and reservoirs. For example, a disinfectant poured down a drain is wasted and needlessly contaminates the environment. If a drain smells, clean it out!

THE ROLE OF ANTIMICROBIAL DRUGS

Antimicrobial drugs do harm as well as good. Some important microbes have learnt how to circumvent them and have appeared as pathogens in hospitals and elsewhere. The mechanisms involved are described as a background to a proper understanding of the purpose and function of the antimicrobial policies discussed in a later chapter.

Box 3.6 Magic pills

As a child, one of the authors suffered from repeated attacks of streptococcal tonsillitis. The first of these was particularly memorable as it was accompanied by scarlet fever and led to isolation in a fever hospital. Later attacks, one or two a year, are remembered as miserable 10-day illnesses with fever, headache, sore throat and, sometimes, acute earache. Treatment was with placebos until one day (probably in 1936) the doctor prescribed some new, large white pills. Their effect was magical: fever, headache and sore throat all subsided in two or three days. The pills were of the first generally available sulphonamide, sulphanilamide. This was soon replaced by the more powerful and broad-spectrum sulphapyridine (M and B 693). Of this drug Winston Churchill, who developed pneumonia in December 1943 while in North Africa, said, 'The admirable M and B . . . was used at the earliest moment . . . and the intruders were repulsed. There is no doubt that pneumonia is a very different illness from what it was before this marvellous drug was discovered'.

Introduction

This section is concerned with the negative aspects of the use of antimicrobial drugs. To balance what would otherwise be an unreasonably gloomy view it is necessary to acknowledge and to stress the contribution these drugs have made to healthcare in the second half of the 20th century. Antimicrobial drugs have produced incalculable benefits, which extend well beyond lives saved and the alleviation of pain and distress (Butler, 1979; Box 3.6).

Substances produced by one type of microbe that can damage or destroy others were called antibiotics when they were first used to treat human infections. Similar drugs made by chemical synthesis were called chemotherapeutic agents and their use was chemotherapy. Since those days many new chemotherapeutic drugs have been produced and a large number of antibiotics have been discovered (Box 3.7), but at the same time the nomenclature has become confused. Some drugs that were originally antibiotics are now produced by chemical synthesis (chloramphenicol is an example), so

Box 3.7 Early antimicrobials

The history of the successful use of chemicals in infections begins in 1619, with the first record of the treatment of malaria with cinchona bark (quinine) in Peru. At about the same time ipecacuanha root (emetine) was found to be useful in the treatment of what is now known to be the amoebic variety of dysentery. Later these alkaloids were joined by the organic arsenicals for syphilis and in 1909 by Erlich's salvarsan for protozoal infections. A major advance came in 1935 when the first true antibacterial agent, prontosil (the original sulphonamide), was introduced. In 1929 Fleming discovered the first antibiotic, penicillin, though this was not used therapeutically until the early 1940s. Today there are over 250 different antimicrobial drugs available worldwide for the treatment of bacterial infections alone, and more that are active against other kinds of microbes.

have changed into chemotherapeutic agents. Ampicillin and other drugs are made by synthetic alteration of an antibiotic precursor, so are hybrids. More recently, a number of compounds introduced for the treatment of cancer have been called chemotherapeutic agents. One solution to this semantic difficulty is to use the term **antimicrobial drug** to cover all the chemotherapeutic agents and antibiotics that are employed in the treatment of infections.

Antimicrobial drugs may be antibacterial, antifungal, antiviral, antiprotozoal and (though worms – helminths – are not microbes) they may include the anthelmintics. All of these are required to act in the presence of two distinct living organisms, to damage or destroy one, and, so far as possible, to leave the other unharmed. To be effective they must reach, combine with and inactivate sets of chemical receptors (targets) sited on or in each of the microbes to be attacked. Each target must be vital to the microbe concerned so that, if it ceases to function, the microbe is crippled or killed. The union between drug and target is firm and highly specific. The image of a lock fitted with a key of exactly the right shape to open it, commonly used to describe the union of an antigen with its antibody, is also appropriate here. When the loss of function is complete the microbe may die (so in the case of bacteria the drug is said to be bactericidal) or it may cease to multiply (with a bacteriostatic agent). Antimicrobial therapy fails if the microbe concerned is of a type that is naturally and permanently resistant to the drug (**natural resistance**) or when a normally sensitive microbe has undergone a reversible change in its chemistry (**acquired resistance**).

Antimicrobial drugs are simultaneously toxic for microbes and non-toxic for the people, animals or plants whose infections they are designed to relieve. This is a tall order and even penicillin, one of the least toxic of these drugs, is poisonous to both parties if enough is given. Other drugs are less safe and for some (the aminoglycosides, for example) the effective antimicrobial dose is uncomfortably close to the level that is toxic for humans.

No antimicrobial drug is poisonous for all microbes. Each antibacterial, antiviral or antiprotozoal drug has a 'spectrum' of activity, which describes the range of bacteria, viruses or protozoa that are damaged or killed by it. Some antimicrobials have a narrow spectrum, so are active against just a few kinds of microbe, while others have a broader spectrum. When treating an infection due to a microbe whose identity is known it is preferable to use a narrow-spectrum drug. This restricts the damage done by 'friendly fire' to beneficial members of the body's normal flora. If such choices are applied coherently by means of an agreed policy the development of antimicrobial resistance can be delayed and, to some extent, kept under control.

Iatrogenic infections are, of course, treated with antimicrobial drugs. Iatrogenic infections differ from others because the patients concerned are usually less able to defend themselves and they may involve hospital strains of bacteria – that is, bacteria that have survived and multiplied in the hospital environment because they are naturally resistant or have acquired resistance to the drugs in common use. When this happens treatment is both more expensive and more difficult. It may involve the use of mixtures of drugs, some of which require the patient's serum to be assayed (tested in the laboratory to determine the amount of drug present) so as to establish the optimum but, at the same time, non-toxic dose.

To have an effect an antimicrobial drug has to surmount many barriers on its journey from the pharmacy to the site of an infection. Even after the chosen drug has arrived and has been administered in the correct dose, by the intended route, at the right time, to a patient for whom it was properly prescribed, other difficulties are encountered. These are of two kinds: first, those imposed by the host and second, those imposed by the microbe itself.

Host factors

It is a common misconception that antimicrobial drugs can, on their own, eliminate the microbes that are the cause of an infection. Antimicrobial drugs are rather weak antiseptics. Though some of them are described as bactericidal, this is a laboratory phenomenon. A total (bactericidal) kill can be achieved in a test tube when small numbers of bacteria are added to a solution of an antimicrobial. If the much larger numbers of bacteria found in real infections are used, many survive. Tests done under these severe but more natural conditions reveal that all antimicrobials are only bacteriostatic. To achieve a bactericidal effect in a real infection the drug must be backed up by the normal antibacterial defences of the body. If these are deficient or absent bacteria survive and, depending on the severity of the deficiency, the treatment is partially or completely unsuccessful.

Clinical experience supports these conclusions. The very few infants born without functional immune systems cannot survive unless they are kept free of microbes by total isolation in a 'plastic bubble' unless or until the underlying cause of their deficiency is corrected by a transplant. Otherwise, despite massive antimicrobial therapy, they succumb to an infection. In AIDS antimicrobial therapy is finally unavailing against the infections that progressively take advantage of the failure of patients' immune systems. Patients prepared for a bone marrow transplant are increasingly likely to suffer from life-threatening infections when their white cell counts fall below $5 \times 10^7/l$. The infections are difficult to control, even with intensive antimicrobial treatment. A more common example is the failure of antimicrobials to clear food poisoning salmonellas from the gastrointestinal tract. Indeed, the usual outcome of their mistaken use in uncomplicated gastroenteritis is further to disturb the normal flora of the gut and to prolong salmonella excretion.

The reason for this is the absence from the gut of the range of most of the body's antibacterial defence mechanisms. Host defences are integral to the clearance of the large populations of bacteria found in natural infections. In patients who recover from infections antimicrobial drugs hasten what, without them, would be the natural process of healing. Antimicrobials save lives and speed recovery, but they do not cure infections on their own.

An antimicrobial that cannot reach its target will fail in its task, even when the patient's immunity is fully intact. Most drugs depend on patients' circulations to carry them to their intended destinations. This cannot happen if the circulation is inadequate. The inadequacy may be due to the infection itself, for example in and around an abscess cavity. In other cases the site of infection may be shielded by fibrosis caused by chronic inflammation or by the encrustations that develop into vegetations in bacterial endocarditis. A restricted circulation is the normal condition in bone, so antimicrobials penetrate poorly into this tissue even when it is healthy.

Some of the most effective barriers to the free passage of antimicrobials are erected by perfectly normal human physiology. Physiological **permeability barriers** keep potentially toxic molecules out of whole organs (in the case of the blood–brain barrier) or operate to prevent the loss of important nutrients from the body (by excretion through the liver and kidneys) or ultimately to preserve the contents of individual cells (Box 3.8). The existence of these barriers limits choices when antimicrobials are selected for the treatment of infections that involve the brain or the biliary or renal tracts. The same applies when the microbes to be attacked multiply inside cells as happens, for example, with *Mycobacterium tuberculosis, Salmonella typhi, Legionella pneumophila* or *Rickettsia* spp.

Box 3.8 Natural permeability barriers

Permeability barriers are important physiological features of all living things. From the single cell of an individual microbe to the mass of them that make up a human body, each is surrounded by a membrane that not only encloses its contents but also monitors and controls what passes in and out. This is how cells capture the nutrients they need, and how they exclude unwanted molecules that might be toxic. (Some toxic molecules manage to gain access, though, if their physical characteristics resemble something more appetizing.) Not only must the membrane allow free passage to useful products and waste materials, but it must also keep vital components locked inside the cell. Larger forms of life erect additional permeability barriers that protect their brains from chemicals harmful to nerve cells that circulate in the rest of the body or in the liver and kidneys to hold back nutrients the body requires at the same time as they excrete waste products.

To be successful, drugs must deceive these guardians of permeability to penetrate the barriers that lie between them and the point where they are required to act.

Microbial factors

After an antimicrobial drug has traversed the hurdles of human fallibility and overcome the problems of permeability in the patient it may meet yet a third set of barriers, in the microbe itself. These are summarized in Box 3.9. The **absence of a target** is typified by the mycoplasmas. These have no cell walls, so are completely resistant to

---

Box 3.9   Mechanisms of microbial resistance

- Total absence of a target for the antimicrobial concerned.
- Lowered permeability that restricts the passage of a drug through a cell wall or a cell membrane, so it cannot reach an otherwise sensitive target within the microbe.
- Modification of the structure of the target (receptor) so the drug can no longer bind to and inactivate it.
- Production of an enzyme that alters the antimicrobial molecule, to make it ineffective.
- The microbe is in a resting phase.

---

all the antimicrobials (the penicillins, cephalosporins and their relatives, plus vancomycin) which, in other bacteria, act to prevent cell wall formation.

The cell walls of many Gram-negative bacteria are more complex than those of Gram-positive ones. Their inner surfaces form a secondary permeability barrier just outside the cell membrane (the primary permeability barrier, found in all cells). This barrier excludes vancomycin which, in consequence, only acts against Gram-positive species. The cell membrane variably excludes some penicillins and cephalosporins, the macrolides and sometimes the aminoglycosides. Many of these drugs would be fully active against targets situated inside the bacteria if only they could reach them, but they fail because of the permeability barrier. Although they are Gram-negative the neisserias are more permeable, so may be fully sensitive to ordinary penicillin, but the less permeable Gram-negative enterobacteria are naturally resistant to it, though they may be sensitive to its relative, ampicillin. Another Gram-negative bacterium, *Pseudomonas aeruginosa*, also excludes ampicillin, so is naturally resistant to it, but it is sensitive to some more specialized penicillins and cephalosporins.

Examples of **structural modifications** are seen in the neisserias and *Streptococcus pneumoniae* that have acquired resistance to penicillin. Receptors called 'penicillin-binding proteins' have been altered. They no longer bind, so are not inactivated by, penicillin. **Inactivation by enzymes** has been mentioned in connection with the production of a beta-lactamase (penicillinase) by *Staphylococcus aureus* (p.38). Other penicillins, together with the cephalosporins, the aminoglycosides and chloramphenicol may also be attacked and inactivated by a variety of specialized bacterial enzymes.

Antimicrobial drugs are at their most effective when they attack microbes while they are multiplying actively. Any large population of microbes includes some that are taking a rest (perhaps in what some have called the dormant or sleepy phase) or, though damaged by the drug, have not been killed. These microbes may begin to multiply again when the drug is removed, either from a test tube or from a patient when treatment ceases.

Development of resistance

Many antimicrobial agents originated as the natural products of microbes themselves. For millennia microbes have used antimicrobials in their struggle to survive and be successful. In any war the appearance of a new weapon on one side leads to the introduction of countermeasures on the other. Some bacteria discovered how to defend themselves when they were attacked by the fungi or other bacteria that produced antimicrobial substances. There was nothing new about antimicrobial drugs when the human race 'invented' them and it explains why 'new' antimicrobial drugs have at best a temporary advantage. The advantage disappears as the protective mechanisms of microbes developed over the centuries are rearranged to deal with an attack, though now from a human source. Perversely, the resistant microbes that then appear are encouraged to multiply as the drug makes space for them by clearing away their more sensitive neighbours.

The development of significant levels of acquired resistance to antimicrobial drugs by clinically important microbes is a constant theme in this book. This is the basis of much of the concern that surrounds IaIs. The acquisition of resistance depends on changes in the structure of the genetic memory (deoxyribonucleic acid, DNA) of the microbe concerned. These changes are inherited and successive generations convert the newly acquired DNA blueprint into one of the mechanisms of resistance listed in Box 3.9.

Bacteria can change their genetic structures in two ways, by mutation or by genetic exchange. Mutations happen by chance and although this is the only way in which entirely novel abilities appear, the new information spreads relatively slowly. The mechanisms of genetic exchange are, by comparison, swift and highly efficient. Short fragments of DNA (preformed genes) are transmitted directly from one bacterium to another. These fragments carry the instructions for some new ability, such as antimicrobial resistance. They spread quickly because they do not have to wait even the short time necessary for bacterial division. Genetic exchanges are more commonly made between bacteria of the same species, but donors and recipients may be of different species of Gram-negative enterobacteria, for example. The fragments of DNA transferred between bacteria in this way may be incorporated either into the bacterial chromosome (nucleoid) or into a structure called a **plasmid**. These are small collections of DNA that exist separately from the main bacterial chromosome, but as they can reproduce themselves they are transmitted from one generation of bacteria to the next.

Transformation, transduction and conjugation

The transmission of preformed genes from one microbe to another is achieved in one of three ways: by transformation, transduction or conjugation (Figure 3.7). In **transformation** naked DNA, released when one bacterium dies and breaks up, is directly absorbed by another. In **transduction** the fragment of DNA hitches a ride as a passenger inside the shell of a bacterial virus (bacteriophage). When

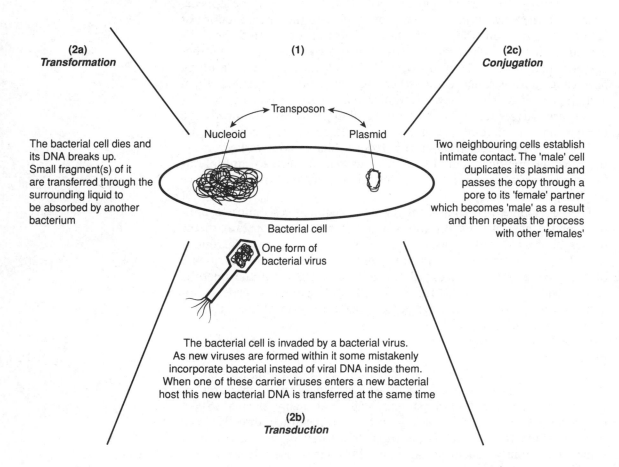

(2a)
Transformation

(1)

(2c)
Conjugation

Transposon

Nucleoid

Plasmid

The bacterial cell dies and its DNA breaks up. Small fragment(s) of it are transferred through the surrounding liquid to be absorbed by another bacterium

Two neighbouring cells establish intimate contact. The 'male' cell duplicates its plasmid and passes the copy through a pore to its 'female' partner which becomes 'male' as a result and then repeats the process with other 'females'

Bacterial cell

One form of bacterial virus

The bacterial cell is invaded by a bacterial virus. As new viruses are formed within it some mistakenly incorporate bacterial instead of viral DNA inside them. When one of these carrier viruses enters a new bacterial host this new bacterial DNA is transferred at the same time

(2b)
Transduction

Figure 3.7 Mechanisms by which pieces of deoxyribonucleic acid (DNA), in the form of individual genes or groups of them, may move (1) *within* a bacterial cell and (2a, b and c) *between* two or more of them. The nucleoid is the main store of the DNA that acts as the blueprint for successive generations of the same type of cell. Plasmids are small accessory stores of DNA. Genes for antimicrobial resistance may be located in the nucleoid or in a plasmid. As illustrated, they can move between these or be transferred to other bacteria to teach a sensitive microbe how to become resistant.

bacteriophages mature inside a bacterial cell some of them acciden-
tally enclose within themselves fragments of host bacterial DNA.
After release from its bacterial parent, the virus attacks another
bacterium and when it has entered it donates the passenger DNA
fragment to a new recipient. **Conjugation** is a quasi-sexual process.
In the course of intimate contact between two bacteria a fragment
of DNA is transferred through a pore that develops between them.
Not only does the recipient acquire whatever instructions are
carried by the fragment, but it also changes its sex to become a
new donor of the same fragment. This highly efficient process
allows a whole population of bacteria to acquire a new piece of
information (for example, how to resist an antimicrobial drug) in a
very short time.

Mutation

Bacterial DNA mutates to produce an inheritable change once in
between 10^5 to 10^{10} cell divisions. Most bacteria multiply rapidly so
after only 24 hours of growth many bacterial populations contain the
progeny of one or more mutations. These mutations proliferate if they
are advantageous to the bacteria that possess them. Clear advantages
arise if a mutation provides a more effective way of utilizing an
existing source of food or if access is gained to new more plentiful
sources or if better methods of neutralizing or avoiding the effects of
poisons are acquired. If a bacterial population that contains a small
number of resistant mutants is exposed to the poison concerned (as
happens with an antimicrobial drug), the sensitive population is wiped
out. The mutants survive, however, and continue to multiply. Their
progeny soon replace the original population, but all are now
resistant. Mutations may emerge in the course of prolonged treatment,
such as are required in tuberculosis and leprosy. This is why two or
three drugs are used simultaneously when treating these diseases.
Mutation to resistance to a single drug is unusual, but not impossible.
The simultaneous or closely spaced sequential occurrence of the
mutations necessary to achieve double or triple resistance is much less
probable.

The emergence of resistance

A method applied in laboratories to establish a resistant strain of a
bacterium is to grow it in a medium that contains a concentration of
the antimicrobial that is a little below the minimum amount at which
multiplication would cease. The presence of the drug does not speed
up the rate of mutation, of course, but it ensures that the desired
mutant, if one does emerge, will survive and declare itself more
rapidly as it multiplies at the expense of its more sensitive relations.

This laboratory experiment may be repeated in the real world, more
particularly when, by accident or design, significant populations of

microbes are exposed to subinhibitory concentrations of an antimicrobial. This happens in a number of circumstances, but perhaps most commonly among patients under treatment for bacterial sexually transmitted diseases. Although not iatrogenic, these infections illustrate the point rather well. The individuals concerned are treated as outpatients so there is a problem with compliance both in the taking of drugs and, for those who make their living from sex, in ensuring sexual abstinence until treatment is complete. The chosen solution to this problem has been to give the treatment in a large single dose, with the intention that the patient is rendered non-infectious as soon as possible. This means that when a resistant mutant appears it continues to multiply after the single dose as the level of antimicrobial in the patient's tissues falls. It is then available in large numbers for transmission to the individual's next sexual partner.

It is instructive to compare the rates at which resistance to antimicrobials have appeared in two members of the genus *Neisseria*, *N. gonorrhoeae* and *N. meningitidis*, the causes of gonorrhoea and meningococcal meningitis respectively. The initially sensitive gonococcus became resistant to the sulphonamides soon after they were introduced, but for some time it remained exceptionally sensitive to penicillin. By 1958, however, strains increasingly resistant to the drug began to appear and it was necessary progressively to raise or otherwise enhance the standard dose used in treatment. In 1976 beta-lactamase- (penicillinase-) producing strains of the gonococcus appeared for the first time and as they spread treatment with penicillin had to be abandoned. Other drugs were substituted and the gonococcus is now resistant to some of these as well.

By comparison, sulphonamide resistance did not became a problem among the meningococci until 1965 and early penicillin resistance was first reported in 1985. This was 25–30 years after the same things had happened to the gonococcus. Beta-lactamase production first appeared among meningococci in 1988, 12 years after the gonococcus. These major differences in the rate of acquisition of resistance by two related bacteria may be explained by the fact that, while gonococci were regularly and deliberately exposed to single-dose antimicrobial therapy, this has not been a feature of the treatment of meningococcal meningitis.

There are ways in which the use of antimicrobial drugs may be controlled, to optimize their benefits, yet restrict their capacity to do harm. 'Antimicrobial policies' are discussed in Chapter 6, p.100.

REFERENCES

Anon, (1988) Splints don't stop colds – surprising! *Lancet*, **i**, 277–8.

Bernander, S., Habraeus, A., Myrback, K. E. *et al.* (1978) Prevalence of hospital-acquired infections in five Swedish hospitals in November 1975. *Scandinavian Journal of Infectious Diseases*, **10**, 66–70.

Butler, C. (1979) Surgery – before and after penicillin. *British Medical Journal*, **ii**, 482–3.

Casewell, M. W. and Phillips, I. (1978) Epidemiological patterns of klebsiella colonization and infection in an intensive care ward. *Journal of Hygiene, Cambridge*, **80**, 295–300.

Chow, A. W., Taylor, P. R., Yoshikawa, T. T. *et al.* (1979) A nosocomial outbreak of infections due to multiply resistant *Proteus mirabilis*: role of intestinal colonization as a major reservoir. *Journal of Infectious Diseases*, **139**, 621–7.

Couch, R. B. (1984) The common cold: control? *Journal of Infectious Diseases*, **150**,167–73.

Cross, A., Allen, J.R., Burke, J. *et al.* (1983) Nosocomial infections due to *Pseudomonas aeruginosa*: review of recent trends. *Reviews of Infectious Diseases*, **5** (Supplement 5), 837–45.

Curie, K., Speller, D. C. E., Simpson, R. A. *et al.* (1978) A hospital epidemic caused by a gentamicin-resistant *Klebsiella aerogenes*. *Journal of Hygiene, Cambridge*, **80**, 115–23.

Edman, J. C., Kovacs, J. A., Masur, H. *et al.* (1988) Ribosomal RNA sequence shows *Pneumocystis carinii* to be a member of the fungi. *Nature*, **334**, 519–22.

Elek, S. D. and Conen, P. E. (1957) The virulence of *Staphylococcus pyogenes* for man. A study of the problems of wound infection. *British Journal of Experimental Pathology*, **38**, 573–86.

Emmerson, A. M., Enstone, J. E., Griffin, M. *et al.* (1996) The second national prevalence survey of infection in hospitals – overview of the results. *Journal of Hospital Infection*, **32**, 175–90.

French, G. L., Cheng, A. and Farrington, M. (1987) Prevalence survey of infection in a Hong Kong hospital using a standard protocol and microcomputer analysis. *Journal of Hospital Infection*, **9**, 132–42.

Glenister, H. M., Taylor, L. J., Bartlett, C. L. R. *et al.* (1992) An 11-month study of infection in wards of a district general hospital. *Journal of Hospital Infection*, **21**, 261–73.

Goldman, D. A. (1992) Epidemiology of *Staphylococcus aureus* and group A streptococci, in *Hospital Infections*, 3rd edn, (eds J. V. Bennett and P. S. Brachman), Little, Brown, Boston, pp. 767–87.

Haley, R. W., Culver, D. H., White, T. W. *et al.* (1985a) The national nosocomial infection rate. A new need for vital statistics. *American Journal of Epidemiology*, **121**, 159–67.

Haley, R. W., Tenney, J. H., Lindsey, J. D. *et al.* (1985b) How frequent are outbreaks of nosocomial infection in community hospitals? *Infection Control*, **6**, 233–6.

Hall, C. B., Douglas, R. G. and Geiman, J. M. (1980) Possible transmission by fomites of respiratory syncytial virus. *Journal of Infectious Diseases*, **141**, 98–102.

Hinton, N. A. (1976) Incidence and character of nosocomial infectious diseases, in *Diagnosis and Management of Gram-Negative Nosocomial Diseases*, (eds I. B. R. Duncan, N. A. Hinton, and J. O. Godden), Medi-Edit, Toronto, pp. 1–12.

Hinton, N. A., Maltman, J. R. and Orr, J. H. (1960) The effect of desiccation on the ability of *Staphylococcus pyogenes* to produce disease in mice. *American Journal of Hygiene*, **72**, 343–50.

Holloway, P. M., Bucknall, R. A. and Denton, G, W. (1986) The effects of sublethal concentrations of chlorhexidine on bacterial pathogenicity. *Journal of Hospital Infection*, **8**, 39–46.

Hovig, B., Lystad, A. and Opsjon, H. (1981) A prevalence survey of infections among hospitalized patients in Norway. *National Institute for Public Health (Oslo) Annals*, **4**, 49–60.

Jepsen, O. B. and Mortensen, N. (1980) Prevalence of nosocomial infections and infection control in Denmark. *Journal of Hospital Infection*, **1**, 237–44.

Kislak, J. W., Eickhoff, T. C. and Finland, M. (1964) Hospital acquired infections and antibiotic usage in the Boston City Hospital – January 1964. *New England Journal of Medicine*, **271**, 834–5.

Leyden, J. J., Marples, R. R. and Kligman, A. M. (1974) *Staphylococcus aureus* in the lesions of atopic dermatitis. *British Journal of Dermatology*, **90**, 525–30.

Maki, D. G., Alvarado, C. J., Hassemer, C. A. *et al.* (1982) Relation of the inanimate environment to endemic nosocomial infection. *New England Journal of Medicine*, **307**, 1562–6.

Martone, W. J., Jarvis, W. R., Culver, D. H. *et al.* (1992) Incidence and nature of endemic and epidemic nosocomial infections, in *Hospital Infections*, 3rd edn, (eds J. V. Bennett and P. S. Brachman), Little, Brown, Boston, pp. 577–96.

McGowan, J. E. and Finland, M. (1974) Infection and antibiotic usage at Boston City Hospital: changes in prevalence during the decade 1964–1973. *Journal of Infectious Diseases*, **129**, 421–8.

Meers, P. D., Ayliffe, G. A. J., Emmerson, A. M. *et al.* (1981) Report on the national survey of infection in hospitals, 1980. *Journal of Hospital Infection*, **2** (Supplement), 1–53.

Meers, P. D. and Leong, K. Y. (1990) The impact of methicillin- and aminoglycoside-resistant *Staphylococcus aureus* on the pattern of hospital-acquired infection in an acute hospital. *Journal of Hospital Infection*, **16**, 231–9.

Meers, P. D. and Yeo, G. A. (1978) Shedding of bacteria and skin squames after handwashing. *Journal of Hygiene, Cambridge*, **81**, 99–105.

Moro, M. L., Stazi, M. A., Marasca, G. *et al.* (1986) National prevalence survey of hospital-acquired infections in Italy, 1983. *Journal of Hospital Infection*, **8**, 72–85.

Noble, W. C. (1983) Microbiology of normal skin, in *Microbial Skin Disease: Its Epidemiology*, Edward Arnold, London, pp. 4–15.

Pitt, T. L., Erdman, Y. J. and Bucher, C. (1980) The epidemiological type identification of *Serratia marcescens* from outbreaks of infection in hospitals. *Journal of Hygiene, Cambridge*, **84**, 269–83.

Rammelkamp, C. H., Mortimer, E. A. and Wolinsky, E. (1964) Transmission of streptococcal and staphylococcal infections. *Annals of Internal Medicine*, **60**, 753–8.

Scheckler, W., Garner, J. S., Kaiser, A. B. *et al.* (1971) Prevalence of infections and antibiotic usage in eight community hospitals, in *Proceedings of the International Conference on Nosocomial Infections*, (eds P. S. Brachman and T. C. Eickhoff), American Hospital Association, Chicago, pp. 299–305.

Selwyn, S. and Ellis, H. (1972) Skin bacteria and skin disinfection reconsidered. *British Medical Journal*, **i**, 136–40.

Shanson, D. C. (1981) Antibiotic-resistant *Staphylococcus aureus*. *Journal of Hospital Infection*, **2**, 11–36.

Sramova, H., Bartanova, A. and Bolek, S. (1988) National prevalence survey of hospital-acquired infections in Czechoslovakia. *Journal of Hospital Infection*, **11**, 328–34.

Wenzel, R. P., Deal, E. C. and Hendley, J. O. (1977) Hospital-acquired viral respiratory illness on a pediatric ward. *Pediatrics*, **60**, 367–71.

Wenzel, R. P., Thompson, R. L., Landry, S. M. *et al.* (1983) Hospital acquired infections in intensive care unit patients: an overview with emphasis on epidemics. *Infection Control*, **4**, 371–5.

FURTHER READING

Bennet, J. V. and Brachman, P. S. (1992) *Hospital Infections*, 3rd edn, Little, Brown, Boston.

Mandell, G. L., Bennett, J. E. and Dolin, R. (1995) *Principles and Practice of Infectious Diseases*, 4th edn, Churchill Livingstone, New York.

Meers, P., Sedgwick, J. and Worsley, M. (1995) *The Microbiology and Epidemiology of Infection for Health Science Students*, Chapman & Hall, London.

Mims, C. A., Playfair, J. H. L., Roitt, I. M. *et al.* (1993) *Medical Microbiology*, Mosby Europe, London.

4 Iatrogenic infections: audit

THE QUALITY OF HEALTHCARE: CLINICAL AUDIT

Audit, what it is, how it is done and how it may be applied to the control of iatrogenic infections. Clinical audit is concerned with the quality of healthcare.

Introduction

Healthcare is no stranger to audit. Apart from straightforward financial control, an early application of *clinical audit* began in 1592, when plague arrived in London. The Company of Parish Clerks, which represented 109 parishes in and near the City, began to publish weekly returns of deaths, with their causes. In 1836 these so-called Bills of Mortality were replaced by the first of today's Registrar General's returns.

Some more recent clinical audits are recorded in Chapters 2, 3, 5 and 6. Many earlier examples were carried out by or for individuals who worked within the organizations audited. They were set up for reasons of personal curiosity or in pursuit of evidence for use in arguments, for example about where to site hospitals. As time passed audits were increasingly imposed from without the organizations they studied, by officials concerned with the outcome of expenditure, measured in terms of benefits produced. Clinical audits may be single events, set to answer single questions. Alternatively audits may be repetitive or continuous and be designed to monitor one or more aspects of the function of an organization, to stimulate those concerned to raise the quantity or quality of their output.

Formal, external audit is fast becoming established as part of the delivery of healthcare. As a parallel development some groups of healthcare workers have established internal systems designed to detect and correct their own errors and inadequacies. Either process is, of course, highly relevant to the control of infection.

The management of quality

Human ingenuity, applied to the manufacture of cars, cameras and colour TVs, can ensure that these are produced without fault, time after time after time. The term 'total quality management' (TQM) has been coined to describe the process by which this is achieved. Internationally agreed vocabularies and standard management systems have been defined.

Can the same thing be achieved in the delivery of healthcare? The human body is infinitely more complex than the most complex machine ever made by human hands. The products of the human reproductive system rather frequently fail to meet the highest specifications. Faults cannot be corrected by recycling the product through the production line. Corrections by medical engineers in later life are comparatively crude and in many cases nothing can be done. Medical engineering is concerned with the whole of human existence, from the womb to the tomb. Until genetic engineering comes of age only a little can be done about any inborn lack of perfection. Most medical effort is directed to the correction of faults after they have become obvious or that arise as the result of an accident. To correct a fault after it has developed is fundamentally different from avoiding it in the first place. Preventive medicine has scored some notable successes, but curative medicine still absorbs most of the time and energy of workers in the healthcare industry. The practice of curative medicine is not free of risk and not all the harm done is avoidable. Patients do not always move smoothly along the healthcare production line and iatrogenic infections are a major contributor to the problems that arise.

Perfection in the delivery of healthcare is unattainable, but the current view is that the application of methods successfully applied to the production of cars, cameras and so on can improve the output of the healthcare industry. The imposition of this idea has caused committees to proliferate and produced thickets of new jargon (Box 4.1). The central concern is with quality. Perhaps to avoid too close a parallel with production lines for cars and to acknowledge a lower starting point, the term 'continuous quality

---

Box 4.1 Burgeoning bureaucracy

The rapid expansion of formal clinical audit is an example of bureaucratic intrusion into the byways of everyday life. For some this is a commentary on the 'nanny' factor in the modern state, for others a realization that many of the individuals in positions of responsibility are unwilling or unable to perform properly unless weighed down with sets of rules, written to cover every eventuality. It might be asked how the world has progressed so far without such things. Molière's M. Jourdain (*Le Bourgeois Gentil-homme*), in discussion with a philosopher, was surprised to discover that he had been speaking in prose for 40 years. The fact is that, just as M. Jourdain used prose, competent managers of old applied the best rules without conscious thought. Not designed by committees, such rules may be more effective than the new, written versions. When all its jargon is rendered down, audit emerges as an exercise in common sense.

improvement' (CQI) has been proposed for application in the health-care field, as an alternative to TQM. The idea that underlies CQI has been explained as a desire to meet the wishes and expectations of patients rather than to support the financial interests of healthcare institutions. In theory, therefore, CQI is not concerned with cost reduction (Simmons and Kritchevsky, 1995). We shall see! A more substantial point about the term CQI is that it contains the important word 'continuous'.

Clinical audit

Quality is monitored by **audit**. When patients are involved in the process the activity is **clinical audit**. As parts of an audit the terms **quality control** and **quality assessment** are used. The output (product) of a unit may consist of objects (including people), words or numbers. Its staff (of whatever size) is responsible for the quantity and quality of their output. Quality control is a process that monitors the qualitative and quantitative perfection of the output, so it is an internal responsibility. Quality assessment (or proficiency testing) is applied by expert assessors from outside a unit, to provide an independent measure of the excellence of the product. Either process may be intermittent or continuous and they are often performed simultaneously. Both are concerned to improve the effectiveness and efficiency of the unit audited. Once established, audit should be a permanent feature. To achieve **accreditation** a unit must meet certain standards laid down by an appropriate, authoritative professional body. The process normally requires, among other things, that the unit submits to audit and the **business plans** drawn up by medical establishments usually include a commitment to it. Audit is intimately connected with **research**. One purpose of research is to define best practice and of audit to see that best practice is applied, in an appropriate fashion.

Accreditation programmes are well advanced in the USA and Australasia. In the USA the Joint Commission on Accreditation of Healthcare Organizations (JCAHO) has identified a series of **quality indicators** that are thought to define the excellence or otherwise of patient care. After 1997 any hospital that seeks accreditation will be required by the JCAHO to collect and report data on these indicators. The intention is to publish the results, so that comparisons can be made. Once more, we shall see. The publication of league tables of performance in healthcare, as in education, has a troubled past. The idea that CQIs are not to be used for financial purposes is less easily maintained when it is learnt that another organization with an interest in this field is called the Health Care Financing Administration! In Australia the JHACO eqivalent is the Australian Council on Healthcare Standards (ACHS).

An audit (or quality audit) may examine the whole output of a unit or be applied to discrete parts of its total function. The performance

---

Box 4.2 Stages in the performance of a clinical audit

- The recognition of a **need.**
- The establishment of the **standard** or **standards** to be applied.
- The **measurement** of performance against these standards.
- A review of the **outcome** of the measurements.
- A review of the need for **change.**
- The definition and **implementation** of the change.

Audit should be a continuous process, so this will lead to:

- **redefinition** of the standards against the performance measured;
- **re-entry** into the audit cycle (above) at the appropriate point.

It may not be possible to establish a standard for a subject not previously examined. In such a case this stage is ignored and the result of the first audit cycle is used to establish a standard for subsequent audits.

---

of an audit may be divided into the stages listed in Box 4.2. Each of these stages is examined in turn in what follows, with special reference to the control of infection.

Need

The stimulus for a clinical audit and the subject of it may arise from official guidance, quality assessment, from an internal review of practice and performance (some form of quality control); as a result of a complaint, request or suggestion; or follow from the publication of the results of research. In many cases the appropriateness of the subject for audit is not an issue, but the choice may be a problem in units whose activities embrace those of others. This is true of infection control, so inappropriate choices are possible. Those employed in this discipline are paid to prevent infections, so the audits they perform should be concerned with counting infections. This can only be done with the active co-operation of other professional groups, whose patients are involved. Quite apart from the delicacy this requires it is also essential that the subject chosen for audit has a clear and direct relevance to infection and its control. A strong attachment to the undeniable aesthetic virtues of tidiness and cleanliness, for example, is not a reason for infection controllers to become involved in an audit process that is the proper concern of domestic managers. The hypothesis that tidiness and cleanliness have something to do with the prevention of infection became prominent over 100 years ago. The continuing lack of scientific support for the idea is highly significant. What evidence there is points in the opposite direction (Maki *et al.*, 1982; pp.51 and 52).

Standards

Standards must be chosen with care. They must not be too high or too low. A standard of 'no infections', for example, is unachievable. Standards need to be set in consultation with those to whom they will apply, otherwise the testing of them against performance is likely to be resisted and the result of the audit ignored. In the absence of anything better, a trawl through the literature may reveal the results of a number of studies of infections and from these provisional standards (**bench marks**) can be set that will become locally relevant on completion of the first audit cycle, when they can be revised to conform to local conditions. Alternatively the first audit cycle may be used to establish the current situation and this is used as the basis for a standard in subsequent audits. When standards are used as targets they become **outcome criteria** or **performance indicators**.

Measurement

The second section of this chapter is concerned with the process of measurement as applied to audit in infection control. For many years the term **surveillance** has been used to describe this activity. Unfortunately this word carries overtones of the totalitarian state and 'big brother'. The now obsolete term **surveyance** might be resurrected to convey the softer, neutral or beneficial concept that is intended. Unwillingness to be involved in appropriate surveyance implies a rejection of audit, with all this means in relation to the current and future management of healthcare and the future of infection control. Those who attempt to justify their existence by reference to no more than an ability to detect and control outbreaks of iatrogenic infections need to be reminded that this ignores the endemic component or 95% of the subject for which they are responsible. Claims to have controlled an outbreak ('We did such and such and it went away') need to be measured against the statement that half of them resolve spontaneously, without intervention of any sort (Wenzel *et al.*, 1983, Haley *et al.*, 1985; p.31).

Outcome, change and implementation

Although implementation can be very difficult, the need for each of the stages named in this heading is self-evident and requires no further explanation. Unless the results of audit are put to good use all the preceding hard work has been wasted.

Redefinition and re-entry

The control of infection is a target that recedes as it is approached, so the quest for it has no end (Chapter 12). This is why audits of IaIs, in particular, have to be continuous. Audit has been described as a circular process in which the end of one cycle acts as a stimulus for the beginning of the next. It might better be thought of as a spiral staircase, in which each new cycle rises to improve upon the previous one (CQI). Successes ought to be documented, and, at the very least, evidence produced that rates of infection are not rising. Another reason for repetitive or continuous audit is that the serial publication of results holds the attention of healthcare workers and motivates them to improve their performance. A word of warning is, however, appropriate. Audit must not be performed to the exclusion of all else. In some places over-enthusiasm for the collection of data by clinical audit has turned it into a mechanical process, performed by individuals who have no time to complete the later stages of the cycle.

Where to audit

Audits of IaIs must bridge the delay (incubation period) between the event that is the cause of an infection and its appearance. IaIs that have been acquired in hospitals increasingly present in the community as the durations of hospital admissions diminish (Chapter 1). Incubation periods vary, but are probably longest in surgical practice. Surgical wound infections are rarely apparent before the third, and they present most numerously between the seventh and 14th postoperative days. Most surgical patients have left hospital by then (see p.148). Various methods have been used to follow patients into the community to avoid the loss of information about any IaIs that develop. These have included home visits and postal or telephone surveys (Zoutman *et al.*, 1990; Ravichandran *et al.*, 1993; Hawkshaw, 1994; Holmes and Readman, 1994; Keeling and Morgan, 1995).

SURVEYS, TRIALS AND STATISTICS

What surveys and trials are, how to arrange them and, as a simple introduction to statistics, how to interpret the results.

Surveys

A **survey** is made as a scientific attempt to investigate the animal, vegetable or mineral properties of a particular location and to describe the characteristics of one or more of the items within it. Surveys may be used to find out how many iatrogenic infections are suffered by patients in hospitals or in other defined locations; what kind of infections are involved; if they are distributed evenly or unevenly within the location surveyed and if the latter, where they are more common; what factors contribute to them; and perhaps to form an opinion as to their cause.

Trials

Trials are two or more surveys linked together to allow comparisons to be made, so as to determine choices. For example, groups of patients may be exposed to different treatments and the outcomes compared to decide which of them is 'the best'. In connection with iatrogenic infections, trials are used to compare different practices or procedures that are or may be applied to control or prevent infections. The practices or procedures may be entirely new or be introduced from elsewhere. New ideas may be compared with each other or with those already in use. Surveys that are linked to form a trial must be carefully controlled. The circumstances and conditions that surround each of them must be as near identical as is possible, with the exception, of course, of the single variable that is the subject of the trial. In a trial the group that is exposed to the new procedure is the experimental or **test group** and the group used for comparison is the **control group**. With these provisos and unless otherwise stated, what follows applies equally to trials and surveys.

Planning

Why survey?

If their results are not put to good use, surveys are an expensive waste of time. When one is contemplated the first action is to define exactly why it is needed. A short-list of reasons for studying iatrogenic infections might include a wish to raise awareness of their existence among members of staff who would otherwise tend to ignore them (as a first step in their control); to identify areas where infections are more common (as a preliminary to a more detailed study aimed at their prevention); to justify expenditure on control measures; to improve performance in their control; and, regretfully, to use the data collected as a defence in litigation. The initial definition is doubly important because the reason for making a survey shapes the decisions that follow.

Type of survey

The next decision is to choose the type of study to be made. The choice is between an incidence and a prevalence approach. A **prevalence survey** involves a count of the number of patients with infections in a defined population at the time of the study. Depending on how many patients are to be included, their geographical distribution and the number of individuals making the survey, it will usually be completed in one or just a few days. As each infection is found its site and type are recorded, together with any other information required. The effect of this is to take a snapshot of the community, frozen at the time of the survey.

An **incidence survey** is similar, but it is conducted over a significant period or it can be continuous. As the population surveyed changes with time, this approach produces a video film compared with the still photograph of a prevalence study. The discharge of patients who develop infections while they are in hospital is delayed, so at any one time individuals with infections occupy a disproportionately large number of beds. A prevalence survey will therefore record more cases of infection than an incidence study, so rates of infection in hospitals, determined by prevalence surveys, are higher than the rates found in incidence studies, in similar populations (see Figure 4.1 for more detail). This does not matter provided the results of the two types of survey are not compared with each other. Incidence studies are the 'gold standard', but they are time consuming, labour intensive and expensive to perform. Prevalence surveys are quick and cheaper to run. Whichever type of survey is chosen and unless it is small, a pilot study involving a few patients should be held to test the methods to be used. This allows any errors or omissions to be corrected before it is too late.

The sample frame

Next, the population to be surveyed must be defined. This sets the **frame** within which the survey is conducted. The frame may enclose patients in a single ward or other small unit, in a whole hospital or in a number of them or patients in some part of the community or any combination of these. Because patients move in and out of hospitals and health centres a frame that includes either of these encloses a population that changes rapidly, which is why incidence and prevalence surveys produce such different results. By comparison the population within a frame based on people's homes is more stable. As a consequence the results of incidence and prevalence studies made in the general community are more similar.

A key factor in planning is to determine the size of the population to be surveyed. This is likely to depend, primarily, on the frequency of the factor to be studied. A prevalence survey of a unit of 20 patients will not discover anything useful about an infection that, on average, afflicts only one in 100 of them. In these circumstances an incidence study of the unit, continued until several hundred patients have passed through it, is likely to be of greater use. For a simple survey of infections in hospitals common sense will be a good guide to the size of the population needed, though it must be noted that more patients are required for prevalence than for incidence studies (Figure 4.1). It should also be noted that, while under-insurance in terms of numbers is likely to be disastrous, over-insurance is not only wasteful but may also introduce an element of confusion.

When a trial is planned the rate of a certain event is to be

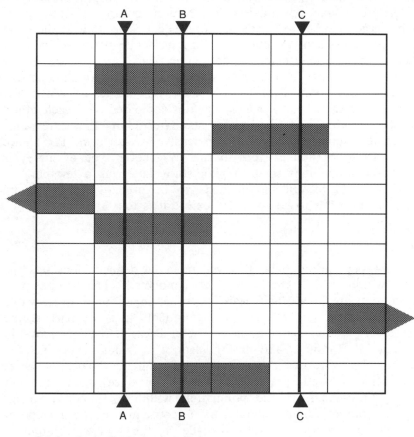

Illustrative example
Defined community (sampling frame): One 12-bedded ward, for 30 days
Average stay of uninfected patients: 5 days (one rectangle)
Average stay of patients with
iatrogenic infections: 10 days (stippled rectangles)

Total patients discharged or died:
• 67, of a possible maximum of 72
The *incidence* of iatrogenic infection:
• 5 patients newly infected, 67 discharged; incidence, 5/67 = 7.5%
The *prevalence* of iatrogenic infection:
• at point A, 2/12 = 17%; at point B, 3/12 =25%; at point C, 1/12 = 8%

N.B. Prevalence rates in hospitals are *higher* than incidence rates. As can also
be seen from this example prevalence results are *unreliable* for small groups,
but they improve as the numbers get larger. Iatrogenic infections are wasteful:
in the example five fewer patients were treated than would have been possible
without them.

Figure 4.1 A comparison of incidence and prevalence surveys in a small
hospital ward.

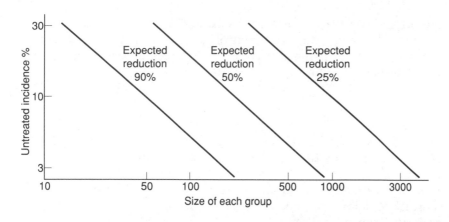

Figure 4.2 The sizes of test and control groups needed to establish a statistically significant result under varying conditions of incidence of the condition under study in the control group and the reduction in this incidence expected as a result of intervention in the test group (for an even chance at the 5% level). See also Box 4.3. (Drawn from data by Lidwell O. M. (1963) Methods of investigation and analysis of results, in Infection in Hospitals, eds R. Williams and R. Shooter, Blackwell, p. 45.)

measured and compared in test and control groups. If the event is common and the procedure to be tested very effective, the groups can be quite small. When Lister introduced antiseptic surgery a mortality rate of 46% in his control group of amputees was reduced to 15% in the antiseptic test group (Lister, 1870). Control and test groups of 35 and 40 were sufficient to prove the point statistically; that is, for the figures to stand up to mathematical analysis using an appropriate statistical test. Figure 4.2 and Box 4.3 show how the sizes of the groups needed to produce significant results grow as the

---

Box 4.3   How many in the trial?

Statisticians can calculate this, but a rough estimate can be made from Figure 4.2. To use this figure it is necessary to:

- know the incidence of the condition to be studied in the control group;
- have an estimate of the reduction in this incidence expected as a result of the intervention proposed.

For example, if a vaccine against an infection that attacks 30% of an unprotected population is expected to reduce that incidence by 90% (to 3%) only about two dozen people, half vaccinated, are needed to produce a statistically significant result. At the other end of the spectrum, to test a measure that is expected to reduce by 25% the incidence of an infection that attacks only 3% of a population requires a trial that involves 6000 people, 3000 in each of the test and control groups.

rate of the event to be measured falls and the expected difference in outcome between test and control groups diminishes. When an important study is planned it is essential to seek expert statistical help to avoid the risk of an expensive mistake.

Criteria and definitions

The last task before detailed planning begins is to choose the criteria or definitions used to describe the infections to be included in the survey. These definitions must be precise to avoid argument later about what exactly was meant, for example, by a wound infection. It is usual to adopt or modify a set of definitions that have already been used somewhere else. This avoids mistakes (or serves to correct any made earlier) and allows results to be compared between surveys. A set of definitions is provided in the Appendix to this book.

The details

More detailed plans will include the identification of the persons who will perform the survey, what additional information they will collect, how the data is to be gathered and recorded and what to do with the results. In the case of a trial it will usually be necessary to secure the informed consent of the patients who will be involved and the approval of an ethical committee or other adjudicator of the propriety of individual medical experiments. The confidentiality of the data will also be an issue for discussion.

The person, or persons, who gather the data must be sufficiently knowledgeable and experienced to have a clear understanding of the nature and purpose of the survey. If it is large and several collectors of data are employed, elaborate precautions are required to ensure conformity between them, so as to secure overall comparability. Computers take some of the hard work out of a survey, but they do no more than that. Unless those who plan a survey or trial are capable of mounting it, in imagination at least, with no more than paper, pencil and a simple calculator, the addition of a computer will not help: indeed, it will only add to the confusion. Another danger of computers is that those involved may be tempted to gather extra data 'in case it might be useful'. Each extra item of data increases the complexity of a survey in geometric rather than arithmetic strides. A watchword for computer users is 'GIGO'. This stands for 'garbage in, garbage out'!

Statistics

At the end of a trial it is usually necessary to apply a statistical test to the results, to determine if the difference observed between test

Box 4.4 Statistical tests of significance

When penicillin first came into use in the early 1940s its effect was so dramatic that there was no need for a large trial to show that it worked. People survived infections that had been lethal or they returned to health in a few days rather than enduring weeks of illness. Such giant strides are rare. Most medical advances require full trials and statistical analysis to prove that they are genuine.

One of the features of the introduction of a novel idea or product, a new drug for example, is that the results of initial trials tend to be more favourable than those that emerge later. One explanation for this is that the first trials are completed under the auspices and close scrutiny of the proud inventor or an anxious manufacturer. Truly independent trials follow, sometimes years later. Unconscious bias can alter the results of trials in which the differences between test and control groups are barely significant. Frank dishonesty in such matters is not unknown!

and control groups is 'significant' or if it might have arisen by chance, due to natural biological variability (Box 4.4). The principle is illustrated by an example. Imagine an experiment in the use of growth hormone in which the variable to be measured is height and the question asked is, does growth hormone stimulate extra growth in normal children and adolescents? For ultimate precision the experiment would have to involve the whole population of young people, with half given growth hormone. This is impractical, so carefully selected samples of the population are used to represent the whole of it. Biological variation determines that average heights measured in successive samples vary somewhat both between samples and from the 'true' figure that would emerge if the whole population was measured. The size of the variation will increase as the number of people in each sample is reduced. To answer the question set it is necessary to compare the average height finally achieved in the test group of young people given the growth hormone with that in the control group, who were not. If the test group turns out to be, on average, a little taller, is this due to the chance variation expected between different samples or is it due to the hormone? Although the question is, in the last resort, unanswerable, statistical tests properly applied give the best approximation to a useful answer that can be made. There are many different statistical tests, each designed for a particular situation. Other than in the most straightforward cases expert help is necessary to make an appropriate choice of test and to ensure that the data is collected in such a way as to be suitable for analysis (and probably to help with the mathematics as well!).

The results of statistical tests may be expressed in different ways One is to give an average with a plus and minus figure added. In the case of height this might be 180 cm +/− 5 cm. This means that if all the results from groups of similar size lie between 175 cm and 185 cm this is as likely to be due to chance as anything else. It follows that if the growth hormone test group produced an average height of 184 cm it would be wise to regard this result as inconclusive and assume that additional growth hormone is without effect on the height achieved by normal people. If the average for the test group was 187 cm, then the opposite conclusion is possible. Results expressed in this way suffer from the defect that they provide no estimate of the degree of confidence carried by the figure: is the deduction made possible, probable or nearly certain? To overcome this uncertainty results may be given in a way that adds this information, for example in the form '95% confidence limits (CL) 175 cm–185 cm'. This gives the same information as the example above but it adds the fact that, if the experiment was repeated 100 times, among the 100 results five would be expected to appear, though still by chance, that give results outside the limits stated. This converts to a single 'rogue' result in 20 examples, 19 of which conform (see later).

A third method is to give the test and control results and to add a figure for the probability that such a result might be due to chance. This is illustrated by an imaginary search for carriers of *Staph. aureus* in successive groups of four nurses taken at random from a large number among whom the overall carriage rate is 50%. On average each sample of four nurses will contain two with positive and two with negative culture results but other ratios are possible and the positives and negatives within each group of four can arise in any order. As testing continues all the 16 variations shown in Table 4.1 will eventually emerge.

From the table it is clear that the **probability** that all the nurses in each group of four might be negative (or positive) is 1/16. The probability of only one of the four being negative (or positive) is 4/16 or 1/4 and that of there being equal numbers negative and positive is 6/16 or 3/8. Note that the fractions add up to 1. The value for the probability (the 'p' value) of absolute certainty is 1 and that of utter impossibility is 0. The nearer the value of 'p' gets

Table 4.1 The results of an imaginary search for carriers of *Staph. aureus*, to show the probabilities (chances) of all the different types of result that are possible, when multiple groups of four individuals are tested, drawn from a large population

Results	*+ or – results, in their order*	*Chances*
No positives	– – – –	1/16
One positive	+ – – –, – + – –, – – + –, – – – +	4/16 (1/4)
Two positives	+ + – –, + – + –, – – + +, – + – +, – + + –, + – – +	6/16 (3/8)
Three positives	+ + + –, + – + +, + + – +, – + + +	4/16 (1/4)
Four positives	+ + + +	1/16

+ denotes a positive and – a negative swab result.

to 1, the more likely it is that an event will take place or that the groups being compared are the same. The closer 'p' is to 0 the more likely it is that it will not or that the groups are significantly different. A p value of 0.5 means that probabilities are 50–50, like those of a spun coin coming down 'heads'. To calculate p from the fractions in the table, the upper figure of each fraction is divided by the lower, giving values of 0.06, 0.25 and 0.38. Some prefer to express chances as percentages, when p = 1 = 100%. The fractions in the table are then 6%, 25% and 38%, respectively.

The value of p that can be accepted as denoting a significant difference lies next to the one that might be due to chance. The choice of figure is arbitrary. The least demanding convention is to say that a difference is significant if a given observation would happen by chance no more than once in 20 otherwise identical trials or less often than that (p = 0.05 or a chance of 5% or less). A more demanding test of significance is to say that the observation would happen by chance not more than once in 100 trials or less often (p = 0.01 or 1% or less). When expressed in the form of confidence limits (CL), a 5% chance becomes a CL of 95% and a chance of 1% a CL of 99%.

This can be illustrated by referring once more to Lister's results. The X^2 ('chi squared') test is simple and is appropriate in this case. Calculation gives a figure that, when applied to published tables of p values, gives a result between 0.005 (0.5%) and 0.001 (0.1%). It therefore emerges that the difference in mortality between control (precarbolic) and test (carbolic) groups as large as Lister observed would be due to chance only once in rather more than 200 trials. The probability that the result of Lister's single experiment was due to luck is thus equivalent to having a coin come down 'heads' eight or nine times in succession. This is highly unlikely so, assuming the groups were otherwise identical, the conclusion is that the improvement was due to the disinfectant.

Among other statistical terms met with in the medical literature are absolute and relative rates, odds and odds ratios and confidence intervals. For enlightenment see, for example, Martin, Plikaytis and Bean (1992).

Application of the results

The results of a trial may be used to validate a new infection control policy or to justify a change in an established one. The replacement of invalidated rituals by scientifically planned policies is to be applauded, but a word of caution is necessary when changes are to be made as a result of studies performed in other places. Iatrogenic infections are a mixture of self, cross and environmental infections (p.44). Call these A, B and C and assume that for a particular type of infection, B (cross-infection) is generally the more common cause. If a trial is done in a hospital in which B is already well controlled, an improvement may be noted if measures are introduced that reduce the influence of factors A or C. Let us assume that when attention is paid to cause C, the incidence of infection falls. The result is published as a success story. When the same measures are applied in another hospital where infections are more common because factor B is still poorly controlled, those concerned will be disappointed. This is because type B infections continue unchecked and their greater number may swamp any small improvement due to the control of factor C. It is not surprising that the results of trials sometimes conflict!

REFERENCES

Haley, R. W., Tenney, J. H., Lindsey, J.O. *et al.* (1985) How frequent are outbreaks of nosocomial infection in community hospitals? *Infection Control*, **6**, 233–6.

Hawkshaw, D. (1994) A day surgery patient telephone follow-up survey. *British Journal of Nursing*, **3**, 348–50.

Holmes, J. and Readman, R. (1994) A study of wound infections following inguinal hernia repair. *Journal of Hospital Infection*, **28**, 153–6.

Keeling, N. J. and Morgan, M. W. E. (1995) Inpatient and post-discharge wound infections in general surgery. *Annals of the Royal College of Surgeons of England*, **77**, 245–7.

Lister, J. (1870) On the effects of the antiseptic system of treatment upon the salubrity of a surgical hospital. *Lancet*, **i**, 4–40.

Maki, D. G., Alvardo, C. J., Hassemer, C. A. *et al.* (1982) Relation of the inanimate hospital environment to endemic nosocomial infection. *New England Journal of Medicine*, **307**, 1562–6.

Martin, S. M., Plikaytis, B. D. and Bean, N. H. (1992) Statistical considerations for analysis of nosocomial infection data, in *Hospital Infections*, 3rd edn, (eds J. V. Bennett and P. S. Brachman), Little, Brown, Boston, pp. 135–59.

Ravichandran, D., Karran, S. E., Toyn, K. *et al.* (1993) Community surveillance: dare we do without it? *Wound Management*, **4**, 68–70.

Simmons, B. P. and Kritchevsky, S. B. (1995) Epidemiologic approaches to quality assessment. *Infection Control and Hospital Epidemiology*, **16**, 101–4.

Wenzel, R. P., Thompson, R. L., Landry, S. M. *et al.* (1983) Hospital acquired infections in intensive care unit patients: an overview with emphasis on epidemics. *Infection Control*, **4**, 371–5.

Zoutman, D., Pearce, P., McKenzie, M. *et al.* (1990) Surgical wound infections occurring in day surgery patients. *American Journal of Infection Control*, **18**, 277–82.

FURTHER READING

Haley, R. W. (1985) Surveillance by objective: a new priority-directed approach to the control of nosocomial infections. *American Journal of Infection Control*, **13**, 78–89.

Nystrom, B. (1992) The role of hospital infection control in the quality system of hospitals. *Journal of Hospital Infection*, **21**, 169–77.

Wardlaw, A. C. (1985) *Practical Statistics for Experimental Biologists*, John Wiley, Chichester.

5 The control of iatrogenic infections

INTRODUCTION AND HISTORY

Something can be done to reduce the contribution made by iatrogenic infections to the risks incurred by people when they submit to medical care. This section describes the background to some of the attempts that have been made, summarizes their outcome and tries to quantify their usefulness.

Introduction

In many countries more people die in hospitals than anywhere else so for this reason alone they are dangerous places. Healthcare is concerned with the natural varieties of morbidity and mortality, but to these are added the results of accidents and the mistakes of healthcare workers. Accidents and mistakes may be physical (falling out of an unfamiliarly high bed, for example, or an overdose of radiation), medical (wrong drug, wrong dose, wrong operation or some other technical error), psychological (reaction to stress) or patients may acquire an infection. Florence Nightingale said that, at the very least, hospitals should do the sick no harm. The nature of life and human fallibility keep her target out of reach, but health professionals need frequent reminders of their duty to strive for safety in the delivery of healthcare.

The beginnings of attempts to control infections in hospitals are lost in history. Hospitals specializing in major community-acquired infections such as smallpox and yellow fever were already in existence at the beginning of the 18th century and venereal diseases were being treated in isolation annexes of hospitals called 'locks'. At the beginning of the 20th century fever hospitals abounded and rules for their management represent early attempts to control infections in hospitals. In most developed countries the number of patients who required admission to these hospitals then began to fall and by the middle of the century they were being closed. Hospitals that specialized in the treatment of tuberculosis were the last to disappear. Most cases of classic infections that still needed special care had to be admitted to ordinary general hospitals, where methods formerly used in fever hospitals were employed to prevent the spread of infection to other patients. These included what was at first called 'barrier nursing' (p.188).

The modern infection control movement

An early suggestion for an ordered approach to the control of hospital-acquired infections (HAIs) appeared in a 1941 memorandum from the UK Medical Research Council. In response to concern about infections in war wounds it recommended that hospitals appoint '. . . full time officers to supervise the control of infection . . .'. In 1944 the same organization proposed that every hospital should set up a committee representing doctors, nurses, laboratory workers and administrators to investigate and to design measures to control all forms of HAI. These separate ideas were brought together at the time of the 1950s staphylococcal pandemic (p.38) and in 1959 hospitals were advised to set up infection control committees (ICCs) as well as to appoint control of infection officers (CIOs). In the same year the last human component was added in Torbay, England, when Miss E. M. Cottrell was appointed as the first infection control nurse (ICN) (Gardner *et al.*, 1962; Meers, 1980). By then the subject of infection control had become a regular topic at scientific meetings and the subject of multiple publications.

These developments were paralleled in other countries and were soon outstripped in the USA, where the 1950s staphylococcus was again the stimulus. In 1968 a training course for ICNs (later called infection control practitioners) was set up and in 1970 the first of a series of International Conferences on Nosocomial Infections was held. The Centers for Disease Control (CDC) provided strong central leadership, backed up by research. Further interest was generated when the American Hospital Association published an infection control manual. The surveillance and control of HAI became an important preoccupation in many US hospitals.

During the 1970s and 1980s there was a proliferation of books, manuals, journals and professional associations concerned with the control of HAI. The Infection Control Nurses' Association (ICNA), set up in the UK in 1970, was joined in 1972 by the Association for Practitioners in Infection Control (APIC) in the USA. The UK-based Hospital Infection Society was established in 1979 and the Society of Hospital Epidemiologists of America was formed in 1980. By that time the study of HAI and its control had acquired the features of a fully fledged medical subspeciality.

The cost of iatrogenic infection

The best estimates of the human and economic impact of HAI (or iatrogenic infection, IaI) come from the USA. Most of the large-scale incidence surveys that have measured all varieties of IaI simultaneously have been organized by the Centers for Disease Control (CDC). The data concerned refer to HAIs because at the time they were collected HAIs were perceived as the principal component of IaIs (p.26). In the late 1960s and early 1970s six (later eight) hospitals were involved in the Comprehensive Hospital Infections Programme (CHIP) (Eickhoff *et al.*, 1969). This detected HAI at the rate of 3.2 infected patients for each 100 discharges. When an adjustment was made for infections that were missed, the rate was raised to 5% (Martone *et al.*, 1992).

This study was followed by the more ambitious National Nosocomial Infections Surveillance (NNIS). This long-term incidence survey has involved some 80 hospitals. Between 1970 and 1982 the NNIS also detected HAI at a rate of 3.2 for every 100 discharges, but these had been counted as individual infections, not infected patients. (A patient who suffers from one HAI not infrequently develops a second or third; pp.30, 31 and Box 3.2). Although the rates measured in the two surveys were the same, the difference in the way they were counted means that the NNIS detected fewer infected patients than the CHIP. Those responsible for the NNIS also adjusted their rate upwards to about 5% to compensate for the incomplete detection of infections (Martone *et al.*, 1992).

The SENIC study

Additional, more concentrated surveys were made in 1970 and in 1975–1976 as parts of the Study on the Efficacy of Nosocomial Infection Control (SENIC). This involved 338 acute hospitals, selected at random. The rate of HAI detected was 5.7 infections or 4.5 infected patients for each 100 admissions. These data were projected to a national incidence of 2.1 million HAIs among 37.7 million patients admitted to 6449 acute hospitals in the USA in a 12-month period, 1975–1976 (Haley *et al.*, 1985). A conservative estimate suggests that about 19 000 deaths were a direct consequence of these HAIs, which may have contributed significantly to another 60 000 (CDC, 1992). These mortality rates may be compared with a projection made from data derived from a study of urinary tract infections (UTIs) associated with urinary catheterization (Platt *et al.*, 1982). This put the annual excess mortality in the USA from this iatrogenic cause alone at 56 000 deaths. Another projection, also made in the USA, suggested that 75 000 deaths a year were associated with bacteraemia (septicaemia) (Maki, 1981). Even if only half of these cases were due to HAI and accepting that virtually all the deaths associated with UTI are septicaemic, these figures suggest that the conclusions about mortality reached in the SENIC study were indeed conservative. It emerges that between one and two patients of every 1000 admitted to hospitals in the USA die of infections acquired there. There is no reason to suppose that this does not apply elsewhere. A survey of 1000 postmortems on patients who died in hospital in Germany concluded that HAIs were directly responsible for the death of 7.4% of them and that in a further 6.3% of cases HAI had contributed to death (Daschner *et al.*, 1978).

Is the control of IaI cost effective?

Some IaI is inescapable, but an unknown proportion of it is avoidable. The SENIC programme was designed to discover if intervention was of value in the control of HAI. It compared the two periods studied (1970 and 1975–1976). In the five-year interval HAI had increased in the hospitals surveyed, overall, by a factor of 10%. When hospitals were grouped according to the effort each had expended in infection control in the interval, significant differences became apparent. Where effort had been minimal an 18% increase in rates of infection was recorded. Where there had been moderate effort, rates had not changed. Where full surveillance of infections at all sites was coupled with active control measures there was a 36% reduction in infections. If all the hospitals surveyed had conformed to the 'best' practice the overall reduction that might have been achieved was calculated at 32%.

On average, a patient with HAI spent an extra four days in hospital. The additional cost of this, projected in 1992 US$, is $2100 for each infection, to reach an annual total of $4.5 billion (Martone *et al.*, 1992). This figure excludes the cost of any associated litigation, and, of course, the pain and anxiety attributable to HAI cannot be quantified. It was estimated that the control programme necessary to produce this result would be self-financing if only 6% of HAIs were prevented (Haley and Garner, 1986; Haley, 1992). A reduction of nearly one-third represents a handsome return on expenditure: in human terms the profit is incalculable. If the SENIC data still apply (Box 5.1) it appears that the scientific application of infection control is highly cost effective. Even more money can be saved if wasteful rituals are also excluded.

More recently, Wenzel (1995) examined the cost effectiveness of a programme designed to control IaI. He based his study on a model 250-bedded hospital that admits 8000 patients a year. It operates a full (US style) infection control programme (p.89) at a total cost of $200 000 a year. Assuming an IaI incidence of 5%, 400 infections are expected. Wenzel applied historical data on the incidences and consequential increases in the durations of admissions imposed by individual infections within this total. This yielded a figure of 1680 extra days spent in hospital. If no more than the additional variable costs of extra days (the 'marginal costs') are considered, the total annual cost at $500 a day is $840 000 or $2100 for each infection. These figures indicate that the infection control programme is self-financing if 95 (24%) of the infections are prevented. The difference between this figure and the 6% quoted above is probably accounted for by an approximate fourfold difference between the marginal and total costs of hospital stay.

Wenzel also used his model to calculate the costs of an infection control programme in relation to the mortality associated with IaI. He took account only of its two most lethal forms: septicaemia and

---

Box 5.1 An old, old story

When describing iatrogenic infections, why is it necessary to use data some of which is two decades old, taken from papers published ten years ago? The reasons are important, and of significance for the future management of infection control.

Hard data are needed to demonstrate the importance of iatrogenic infections and that their control is worthwhile. They must be presented in a form that engages the attention of workers, administrators and financial managers in the healthcare industry and, if need be, members of the public as well. To do this the data must underline the importance of the whole range of iatrogenic infections or at least all of its hospital-acquired component. The SENIC study did this, so the data it yielded have been, and are still used to make extrapolations to describe the human and economic burdens imposed by iatrogenic (or hospital-acquired) infections.

Such data can only be collected by incidence surveys of an adequate size. Incidence studies are comparatively expensive to perform so should not be too big. Those involved must be dedicated to the task, and be prepared to work hard at it. It may be that the SENIC study was too large and so unnecessarily expensive. This, together with some complacency, may be why others have been deterred from attempting a similar exercise. New, carefully designed studies of iatrogenic infections are long overdue. Data are badly needed to show that iatrogenic infections and their control are (or are not) still relevant, and that the US experience can (or cannot) be applied in other countries. Sooner or later infection controllers will have to justify their existence. They must show that there are infections to be controlled, that it is possible to control them and that they, personally, can contribute to the process.

---

pneumonia. From historical data he assumed an excess mortality due to these two infections of ten and four, respectively, to give a minimum total of 14 deaths directly attributable to IaI. If 20% of these infections were prevented, about three lives are spared each year. To quantify the value of these lives is to venture onto thin ice, but health economists have introduced the 'quality of life year' (QALY) in an attempt to do this. Assuming 20 further years of life in excellent health, three lives spared is 60 QALYs or 30 if the quality of life is reduced to 50% of 'normal'. If the cost of the control programme is divided by these figures the cost of each QALY lies between $3300 and $6600. Wenzel compared this with estimates for other healthcare interventions. Coronary artery bypass grafts emerged at $5100 for each QALY, neonatal intensive care of infants of 500–999 g birthweight at $38 580, continuous ambulatory peritoneal dialysis at $57 100 and a liver transplant at $250 000. Once more, the cost of an effective IaI control programme emerges as money well spent.

ORGANIZATION

Various systems have been devised for the organization of iatrogenic infection control. Some of their similarities and differences are described and the functions of the infection control team, usually made up of a doctor (part time) and a nurse (full time), are outlined.

Personnel and their skills

The basic skills that underlie any attempt to control iatrogenic infections are a sound understanding of the diagnosis, treatment, microbiology and epidemiology of infectious diseases, in the special setting of the delivery of healthcare. An intimate knowledge of the geography and administrative structure of the medical unit concerned (either in hospitals or the community) is also necessary, plus an ability to communicate effectively with staff at all levels, individually or in groups. The availability of time to do the job and the patience to see it through are final prerequisites. The skills and talents listed may be found in medically qualified microbiologists, infectious disease physicians or epidemiologists, who need to have received additional specialist training in the subject of IaI. Such people may be appointed as control of infection officers (CIOs), control of infection doctors or infection control physicians (CIDs or ICPs) or whatever title is appropriate. Because such individuals are employed primarily to perform other duties, one or more infection control nurses (ICNs), nurse epidemiologists, infection control practitioners or an equivalent title, who have also received specialist training, may be appointed to support them. The combination forms a small but potentially highly effective **infection control team** (ICT).

Administration

In hospitals an ICT normally reports to and is monitored by an **infection control committee** (ICC). This committee should have among its members senior administrative personnel from all the main departments of the hospital, plus influential clinicians from the major specialities. The chairman should be a respected senior member of the hospital staff who has executive authority. Unless a CIO possesses this authority they should not be chairman: indeed, it can be argued that such appointments are flawed because CIOs cannot effectively monitor their own activities. The committee needs to meet regularly to support a newly formed ICT or when for some reason an established team is in difficulty. An experienced and effective team needs to be monitored less often and the committee may then only meet to receive reports, deal with large problems or endorse major policy decisions.

As the importance of the community in the control of IaI has become more obvious (Chapter 1), many ICCs have broadened their responsibilities and have included representatives from community medical (public health) services. Such cross-representation is essential, but other administrative structures may be more appropriate in different circumstances. Whatever solution is adopted must take account of the requirement to co-ordinate the hospital with the community aspects of the control of IaIs.

Functions

To be effective an ICT must fulfil the functions summarized in Table 5.1 under the headings input, digestion and output. Input is of two kinds: routine and occasional. Routine input is the regular accumulation of information on infections present in a unit (that is, audit). This is an active process if it involves one or more members of the team in a direct search for cases of infection or passive if notifications result from the actions of others. These approaches are described below. Occasional input consists of individual questions on matters concerned with the control of infection directed to members of the team or chance observations made by them. Some queries require immediate responses – for example, what to do about a patient with an unexpected infection problem or the action to take when a vital machine breaks down. Other questions are more general, as might arise from the introduction of a new fibre optic diagnostic or therapeutic tool that requires modification of an existing sterilization and disinfection policy. Another example is the contribution made by an infection control team to plans for the establishment of new intensive care, oncology, transplant or other high-risk units. Enquiries of these kinds may come from medical or nursing staff or from members of the support services. Of the latter administration, central sterile supply, pharmacy, catering, domestic, laundry and engineering services provide the largest number of queries, but an effective ICT eventually establishes contact with everyone concerned with the delivery of healthcare. A courteous and helpful response to queries, even if repetitive and seemingly pointless, is the hallmark of a good team.

Members of an experienced ICT are able to respond immediately to

Table 5.1 A flow-chart to summarize the functions of an infection control team

---

Input
Routine
   Information on infections, collected continuously, either *actively* or *passively*, usually by means of surveys (surveillance, audit).
Occasional
   Queries relating to acute problems or policy.

Digestion
Use of the input (above) to monitor the level of IaI and, when necessary and appropriate, to plan corrective action. The formulation of responses to queries. The identification of other problems and the design of solutions.

Output
The delivery of advice to individuals or groups, orally or in writing, to include the preparation of policies and procedures. Preliminary and continuing education of all grades of staff. Publication and dissemination of the results of audits.

---

many of the questions they are asked because they have met and dealt with the same problem before. New problems require 'digestion' while members of the ICT discuss and agree a solution. Without consensus the output of a team will be inconsistent, lack credibility and will, quite properly, be ignored. The methods employed to reach decisions vary with the knowledge and experience of team members, but this book and others like it are designed to contribute to the process.

As described in Chapter 4, p.75, before any kind of survey (audit) can begin each variety of infection to be counted must be defined. Examples of definitions are given in the Appendix. Definitions ought to be simple and memorable. Over-sophistication is self-defeating, particularly where infection control activity is poorly developed and when returns of infections are completed by non-specialist staff.

At one time many ICTs involved themselves in a significant amount of bacteriological monitoring of the environment. Floors, walls, the air, water, food, medications and disinfectants might all be sampled on a routine basis and sterilizers checked for their ability to kill bacterial spores. Most of these activities have been abandoned as unhelpful, other than when clearly indicated as a part of the investigation of an outbreak of infection. Routine checks may still be carried out (perhaps weekly) on infant feeds prepared in hospital milk kitchens or on hospital water supplies if these are of doubtful quality. The testing of sterilizers is discussed in Chapter 6, p.113.

Nearly complete data collection

In the USA a national system for the routine collection of input data was developed, based on systematic active surveillance. In its complete form ICNs (nurse epidemiologists, infection control practitioners) were required to spend a significant amount of time searching for infections. Every day the results of cultures sent to microbiology laboratories were scrutinized and X-ray departments and wards were visited. This was to identify as many patients as possible whose cultures or films suggested the presence of an infection or whose records indicated a fever, the use of antimicrobial drugs or the application of isolation precautions. Records were kept and periodic summaries produced and circulated. A continuous incidence survey of this sort (p.73) occupied one full-time nurse with clerical support for every 250 beds served. This expensive approach can be justified if something useful is done with the data collected.

Less complete data collection

Less expensive options were developed elsewhere. In countries whose clinical microbiology services are rudimentary or absent, other sources of routine input data must be found if anything at all is to be done. The substitute can take the form of regular, usually weekly, returns made by the staff of wards or clinical departments in hospitals and of medical units in the community. Cases of IaI that have been noted are listed. The standard form designed for this purpose should not require so much detail that this alone deters those who should complete it. Information derived in this way is usually of poor quality, but it provides a crude measure of events when records from each area are compared week by week and in any case it is better than nothing. The task of data collection is, of course, subordinate to the primary responsibilities of the individuals who perform the task. In general, non-specialists have a poor understanding of the nature and extent of IaI and they tend to play down the much larger endemic component of it (p.31). The rather weak system just described can be improved by the recruitment of designated individuals (often nurses) from each of the 'units' identified for the collection of data, who are given rudimentary training to motivate them and to improve their performance. Such individuals have sometimes been called **link** (or **liaison**) **nurses**.

Link (or liaison) nurses may also be appointed to improve the performance and geographical reach of more active ICTs. They can be used to collect the data required for various forms of audit (p.68). This creates a large number of extra data collection points and data collectors, whose output must be standardized to ensure the comparability of the information they gather. There is a possibility that collectors, whose primary loyalty lies with the unit in which they work, will (even subconsciously) suppress data that might be interpreted as critical of their unit. Independent monitoring and validation are essential and must be applied in a manner agreed by all concerned, including the ICC responsible for the programme.

The collection of IaI data in this way is a passive process, at least so far as the ICT is concerned. Microbiology laboratories are an alternative and better source of passively derived information. To be fully effective in this respect the laboratory must fulfil three criteria. It must process a significant volume of microbiological samples, received from a representative selection of the patients in the unit concerned and the laboratory staff must include one or more people with the requisite knowledge and commitment to identify and notify potentially significant results. Such individuals must possess the skills listed in the first sentence of this section. In addition a human 'sixth sense', developed by experience, is required if a significant proportion of endemic IaI is to be picked up in addition to the outbreaks detected by the much simpler 'alert organism' approach, described below. This human 'sixth sense' cannot easily be duplicated by a computer programme.

The full process is as follows. Every culture result that suggests a relevant infection is identified and the ICN (or equivalent) is informed. Ideally they then visit the patient from whom the specimen came or examine their notes. If this is impossible, the necessary contact is made by other means. The purpose of this is to decide, *in each case*, first, if the culture relates to a colonization or an infection (p.25) and, second, when the presence of an infection has been confirmed, to distinguish between IaIs and non-iatrogenic infection of community origin (community infections). Without this validation further activity is a waste of time. Cases identified as IaIs and significant colonizations are investigated further and other relevant details are recorded. At the same time the opportunity is taken to talk to the staff of the unit concerned to discover any related or unrelated infectious problems and to disseminate information. When visits are complete the ICN returns to base (ideally in or near the laboratory) to log the information collected.

This method of surveillance was named 'laboratory-based ward liaison surveillance' (LBWLS) in a useful booklet (Glenister *et al.*, 1992), where it was described as the 'most effective' of eight incomplete surveillance methods that were identified by the authors. Information derived from only one source is, of necessity, incomplete. When laboratory reports are used the proportion of IaI detected depends on the frequency with which suspect or clinically diagnosed infections are sampled microbiologically and are identified in the laboratory as potential examples of IaI. The number of IaIs identified because specimens are taken as a proportion of their total number is unknown, so the real incidence of IaI can only be guessed at. This imposes a need for great caution if the data collected in this way are to be used to compare the performance of different hospitals or medical units.

The rate of sampling may approach 100% in, for example, intensive care units but in other areas the rate is not only lower, but it also varies widely between different units and departments. In the NUH (p.32) it was estimated that, over the whole hospital, about 50% of all infections were detected by this technique. Glenister *et al.* claimed a rate of 70%, but their survey excluded some clinical services. Whatever the scale of the deficiency, information collected carefully and consistently gives data that are comparable over months and years with regard to the proportion of all IaI sampled, its types and distribution and its microbial causes, *but only when applied within the unit surveyed.* Any changes noted call for investigation. An experienced ICN properly integrated into a microbiology laboratory can perform this kind of incomplete incidence study for up to 600 beds, with time to spare for other infection control duties.

Other forms of less complete incidence study are used or can be imagined. One method has been the so-called **'alert organism'** approach. In this case certain microbes are listed and a report of the isolation of one of them triggers a response (similar to that just described) by the ICT. 'Alert' organisms are usually defined in the same way as 'hospital strains' to include multiresistant bacteria (p.41). The microbes listed are those more likely to be associated with outbreaks of infection and do not include those that cause the great mass of endemic infection, that is, some 95% of the whole of IaI (p.32). The detection of outbreaks is socially and politically important, but to exclude the major part of what is their proper concern is a poor advertisement for infection control organizations that depend wholly on this approach.

In other forms of surveillance intensive care units might be subjected to continuous complete surveys, with sporadic activity directed at other clinical areas, in rotation. The results may be used to detect changes in the incidence of one or other kind of IaI or to monitor the effect of new policies or to discover how and why antimicrobial drugs are being prescribed. They can also be used to impress hospital staff with the importance of IaI. While this is going on a continuous watch is maintained for evidence of unusual infections anywhere in the medical establishment concerned. This duty falls primarily to the person who signs or otherwise validates laboratory reports before they are released so it, together with the 'alert organism' approach, represents a rudimentary form of the LBWLS described above.

Displaying the information

An easy way to record and display the information collected is to use 'T' cards and a slotted board, as illustrated in Figure 5.1. This method is described to illustrate the steps essential to any system used for this purpose. Although deceptively simple, 'T' cards make the detection of clusters of similar infections or of infections due to the same microbe a matter of simple observation. Information can be updated daily in a way that relates microbial pathogens, sites of infection and the clinical areas involved in a visually easily assimilated fashion, instantly accessible to all those who need to know about it. If the display is made in the laboratory the technical staff are included and their output is focused and made more relevant. At the end of the month cards that refer to patients who have been discharged are removed from the board. The information they carry is stored in whatever format is preferred; at the simplest level they are filed in a box.

Computers can replace 'T' cards and boards. The strength of a computer is that periodic statements can be extracted and special surveys made with minimum additional effort. Its weakness is that it cannot recognize an outbreak or trigger any corrective action in a timely fashion unless both data entry and full file interrogation are performed daily. Computer printouts of raw data require knowledgeable interpretation. Some people find a stack of them incomprehensible, so the information may not be available to all who need it. Periodic extraction of information from 'T' cards into a more compact form (including though not necessarily by entry into a computer) yields data that can be converted into a continuous incidence study of IaI. The combination of a 'T' card system for daily use with a computer to assimilate the data on (say) a monthly basis would be a

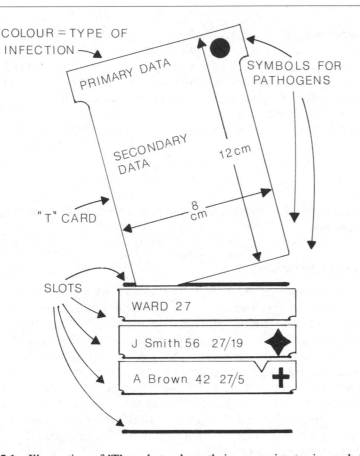

Figure 5.1 Illustration of 'T' cards to show their approximate size and shape. These can be cut out of stiff card if the commercially available variety cannot be found. The colour of the card (four or more different colours may be used) indicates the type of infection, say yellow for urinary, red for surgical wounds, etc. Alternatively a single colour may be used with coloured self-adhesive flashes to indicate the infection. Similar flashes of different colours and shapes indicate the six or seven most common pathogens. The primary data describe the patient and their location, the secondary will include such things as dates of admission and eventually of discharge, clinical diagnoses, antibiograms of pathogens, dates of surgery, catheterization and so on, according to the wishes of the ICT. Cases of significant CIs may be included, particularly if transmission is possible within the hospital, to help identify any subsequent spread. To prevent confusion between CIs and IaIs, a notch may be cut in the top of cards that record CIs (see the lowest card in the illustration). The slotted board used to hold the cards, made of pressed steel sections, is available from suppliers of office equipment. Enough of these sections are joined together so that each ward or other clinical area is allocated its own set of slots: eight to ten usually suffice.

very effective solution. The periodic computer printouts can be refined so as to be comprehensible to those who will receive and might benefit from them.

It must be stressed that anyone who intends to collect IaI data but

who does not fully comprehend how this can be done by hand, using 'T' cards or in some other simple way, is most unlikely to make proper and intelligent use of a computer programme, no matter how apparently sophisticated or oversold it may be. Data must be validated before they are entered into a computer and if this vital step is omitted the condition of 'garbage in, garbage out' (GIGO) will apply. Computers are not a substitute for 'shoe-leather' epidemiology.

DELIVERY

The output of an infection control team is described, as the final stage of its total function.

Day-to-day operations

Information emerges from ICTs as *advice*. CIOs and ICNs should rarely if ever be given executive authority. Final authority in clinical matters belongs to those with full responsibility for patients and this cannot be usurped or shared without threatening their welfare. More practically, there can be no output from an ICT that lacks input. Potential contributors of potentially valuable input are likely to withhold it if they cannot exert some control over the actions that might follow. It is important to remember that advice concerning infection control is only one of a number of often competing and perhaps conflicting factors that must be considered in reaching decisions in medical establishments.

The output from an effective ICT ranges over nearly all aspects of healthcare activity. It is likely to include contributions to long-term building and operational plans, the use of disinfectants and antimicrobials and the management of areas with particular infection problems such as dialysis, transplant, oncology and intensive care units. Apart from the more obvious areas such as patient isolation, the testing of autoclaves and the safety of food, other topics with which ICTs are involved include the disposal of waste, methods used for cleaning and the handling of laundry.

Much advice may be consolidated into written procedures and policies (Table 5.2). It would be possible to divide all hospital activity into a large number of concise, individual procedures, each recorded step-by-step. The result would be voluminous, repetitious, confusing and difficult to co-ordinate. It would almost certainly be ignored and so wasted. A limited number of discrete activities in the field of nursing are, however, best handled in this way. Infection control input is essential, for example, when procedures are formulated for dressing wounds or the management of urinary catheterization, tracheal intubation or venous access. In other cases advice is best incorporated into policies. These describe the basis for certain activities that are related by common factors such as the choice and use of antimicrobials, the

Table 5.2 An outline of the elements that make up a complete service designed to deliver a programme for the control of iatrogenic infections (see also Table 5.1)

Set up an infection control team and an infection control committee to monitor it. The team:

- *conducts* a programme of surveillance (audit)
- *advises* other healthcare workers, as required
- *reacts* to problems related to infections, as they arise
- *formulates* policies (antimicrobial, disinfection & sterilization, isolation, laundry, waste, staff vaccination and safety, etc.)
- *contributes* to the design of procedures that are applied when using catheters (urinary and IV) and ventilators or when performing dressings, etc.
- *monitors* the provision of adequate facilities for handwashing, sterile supply, isolation and in operating departments, kitchens, etc.

provision of an economical set of multipurpose disinfectants, the need for and methods to be used in patient isolation, how to deal with laundry or handle waste. A disinfectant policy is likely to include everything from the preparation of the skin for a variety of different procedures to the method used for cleaning and disinfecting endoscopes.

Each independent unit or hospital should decide which activities they wish to co-ordinate and how these should be divided between procedures and policies. The necessary administrative structures are then devised to ensure that everyone who should be, is consulted. This is important to secure compliance. For example, when writing a new procedure the nursing procedures committee (or equivalent) should seek advice from the ICN (or from the ICT or ICC through the ICN) on matters relating to the control of infection. For activities that involve wider sections of the healthcare community the ICC or some other appropriate body may set up subcommittees or small *ad hoc* bodies to devise the necessary policies. These should have powers of co-option so, as an obvious example, an endoscopist is involved when the cleaning and disinfection of endoscopes is discussed.

Information is disseminated in a variety of ways. To be effective ICTs have to develop lines of communication with more categories of healthcare workers, individually or in groups, than anyone else, senior administrators excepted. An effective team must be able to identify, reach and influence those who make executive decisions, from the least to the most important. This involves many visits, much talking and a lot of shoe-leather.

Most acute questions are answered face-to-face or by telephone. Less acute matters and policies are handled differently and the method will vary with the administrative structure involved. When administrative power is vested in a single individual or a small group, advice is best channelled through an ICC, particularly if the chairman is a powerful figure. In units in which power is more diffuse, committees proliferate and user bodies may have considerable influence. Most of these groups will at some time hold discussions that impinge on infection and its control. An ICT should be prepared for its members to be co-opted to such committees, as required, to help make cost-effective decisions and prevent mistakes.

Education

An important part of advice giving is education. This may be used in two ways. The publication of infection surveillance data ought to educate those who receive it. Unfortunately even when it is produced it all too often ends up unread in the wastepaper basket. One method that has been used to attract attention is to publish it so that recipients can recognize their own contribution to IaI without it being obvious to anyone else. This approach has reduced the incidence of surgical wound infections (p.141). Another application of education is to make staff aware of the existence of IaI and to impress on them their duty to control it. This is best done during the formal education of doctors and nurses, with periodic in-service refresher courses. Other healthcare workers should be taught the principles of infection control as applied to their own work when they are appointed, again with refreshers from time to time. The elements of a complete infection control programme are summarized in Table 5.2 (see also Table 5.1).

Throughout their careers most doctors spend more time dealing with infections than with any other variety of organic disease. It is surprising that, in general, medical schools fail to educate medical undergraduates in a subject that will occupy such a high proportion of their working lives. As a coherent topic, IaI may not even be mentioned in undergraduate courses. Although nurses are taught about the prevention of infections, tutors in schools of nursing are unlikely to have received adequate formal or any continuing education in the developing science of iatrogenic infection control. Some recognize this and use a qualified ICN to make up the deficiency. Others, unfortunately, do not. The result is that some student nurses are still taught wasteful and illogical rituals, based on defective ideas current as much as 50 years ago.

The need for a formal programme to monitor and control IaI is, in the final analysis, attributable to these failures of education. The delivery of healthcare to individual patients lies in the hands of individual healthcare workers. The safety of the delivery, and particularly the avoidance of infection, is determined at the same level. The requirement that doctors and nurses should be monitored in this part of their overall work is a measure of the seriousness of this failure. The problem needs to be tackled at its roots. The object should be to make ICNs redundant or, more positively, to make all healthcare workers their own ICNs. In the developed parts of the world inertia in the existing system makes this a distant prospect. Paradoxically it might more easily be achieved in less developed areas. Where infection control programmes are themselves only ideas, a relatively minor expenditure on primary and continuing education would have a disproportionately large effect.

The average physician or surgeon tends to have a heavy commitment to the patient of the moment. Other than by epidemiologists, little thought is given to patients in the mass. Nurses, though also oriented towards individual patients, are trained to care for them in groups. Although infection control rests on an intimate relationship between each patient and a variety of healthcare workers, Table 5.2 shows that individual infection control behaviour is determined by group decisions. Not only are nurses more likely to react positively to these, but they are more numerous and so come into intimate contact with patients more often than other healthcare workers. Of course, all healthcare workers need to be better educated, but this analysis suggests that effort directed to nurses will produce the greatest return most quickly. The points at which maximum benefit can result from minimum input are in medical schools and schools of nursing, with the emphasis on the latter. Education of the educators is the key and as with all wisdom, the recognition of ignorance is the first step.

REFERENCES

CDC Hospital Infections Program (1992) Surveillance, prevention and control of nosocomial infections. *Morbidity and Mortality Weekly Report*, **41**, 783–7.

Daschner, F., Nadjem, H., Langmaack, H. *et al.* (1978) Surveillance, prevention and control of hospital-acquired infections. III. Nosocomial infections as cause of death: retrospective analysis of 1000 autopsy reports. *Infection*, **6**, 261–5.

Eickhoff, T. C., Brachman, P. S., Bennett, J. V. *et al.* (1969) Surveillance of nosocomial infection in community hospitals. 1. Surveillance methods, effectiveness and initial results. *Journal of Infectious Diseases*, **120**, 305–17.

Gardner, A. M. N., Stamp, M., Bowgen, J. A. *et al.* (1962) The infection control sister. *Lancet*, **ii**, 710–11.

Glenister, H. M., Taylor, L. J., Bartlett, C. L. R. *et al.* (1992) An 11-month study of infection in wards of a district general hospital. *Journal of Hospital Infection*, **21**, 261–73.

Haley, R. W. (1992) The development of infection surveillance and control programs, in *Hospital Infections*, 3rd edn, (eds J. V. Bennett and P. S. Brachman), Little, Brown, Boston, pp. 63–77.

Haley, R. W. and Garner, J. S. (1986) Infection surveillance and control programs, in *Hospital Infections*, 2nd edn, (eds J. V. Bennett and P. S. Brachman), Little, Brown, Boston, pp. 39–50.

Haley, R. W., Culver, D. H., White, J. W. *et al.* (1985) The national nosocomial infection rate. A new need for vital statistics. *American Journal of Epidemiology*, **121**, 159–67.

Maki, D. G. (1981) Nosocomial bacteraemia: an epidemiologic overview. *American Journal of Medicine*, **305**, 731–5.

Martone, W. J., Jarvis, W. R., Culver, D. H. *et al.* (1992) Incidence and nature of endemic and epidemic nosocomial infections, in *Hospital Infections*, 3rd edn, (eds J. V. Bennett and P. S. Brachman), Little, Brown, Boston, pp. 577–96.

Meers, P. D. (1980) The organisation of infection control in hospitals. *Journal of Hospital Infection*, **1**, 187–91.

Platt, R., Polk, B. F., Murdock, B. *et al.* (1982) Mortality associated with nosocomial urinary-tract infection. *New England Journal of Medicine*, **307**, 637–42.

Wenzel, R. P. (1995) The economics of nosocomial infections. *Journal of Hospital Infection*, **31**, 79–87.

FURTHER READING

Bennet, J. V. and Brachman, P. S. (1992) *Hospital Infections*, 3rd edn, Little, Brown, Boston.

Mandell, G. L., Bennett, J. E. and Dolin, R. (1995) *Principles and Practice of Infectious Diseases*, 4th edn, Churchill Livingstone, New York.

Disabling or killing microbes

INTRODUCTION

This section introduces ideas central to the disabling and killing of microbes as an introduction to the major topics dealt with in the rest of Chapter 6.

Microbes are the most numerous living things on earth. Nearly all of them function to break down the bodies of plants and animals after they have died, to release vital nutrients that are used again. Only a few are able to attack their larger neighbours while they are still alive, to cause disease. There would be no infections if all the microbes were destroyed of course, but as nutrients would not be recycled either there would soon be nothing left alive to suffer from them! Microbes should not be killed indiscriminately, but selectively and only when it is necessary to inactivate or destroy those that might do harm. This is not easy. Microbes share with humans, animals and plants many of the chemical and biological processes that constitute 'life'. The most effective ways of killing microbes are, therefore, also lethal for everything else. Each of us is home to many millions of microbes. We have learnt to coexist and we even use some of them to defend us against other microbes (p.22). When we need to attack microbes in or on us the method must be chosen with care to avoid damage to ourselves and, if possible, to preserve those that are useful.

So far smallpox is the only infection that has been eradicated. Vaccination was the key to success and in the future other vaccines may make it possible to eliminate such infections as poliomyelitis, measles, mumps, chickenpox and hepatitis. Simple coarse filtration of drinking water has helped to bring the Guinea worm (*Dracunculus medinensis*) to the brink of extinction, but it is impossible to visualize the disappearance of, for example, such major iatrogenic pathogens as the staphylococci and streptococci. Microbes adapt to new conditions more cleverly and much more quickly than humans, so they will survive long after we have disappeared.

The word 'disabling' is included in the title of this chapter for two reasons. First, viruses are not alive so cannot be killed, but when they are disabled they no longer multiply to cause infections. Second, some useful antimicrobial drugs do no more than stop bacteria

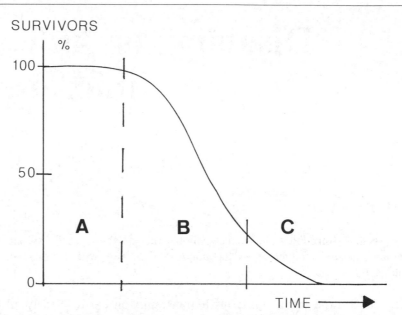

Figure 6.1 The result of applying a disabling or killing agent to a population of microbes, at the beginning of period A. During the remainder of this period the agent reaches its target and begins to have an effect. During period B killing or disabling begins and progresses at a steady rate and in period C more resistant or inaccessible microbes are dealt with. The whole of the period A + B + C may measure minutes, hours or last for ever, depending on the power of the agent and the resistance of the microbes.

multiplying (p.55). When the destruction of microbes is discussed another difficulty has to be accommodated. In the human context the words 'alive' or 'dead' are nearly always applied to individuals, one at a time. Microbes, on the other hand, must be thought of in whole populations, numbered in millions. Even if 99% of a million germs are killed, 10 000 are left alive. A few hours later, under favourable conditions, this residue can again number a million.

As happens with any form of life gathered together in large numbers, populations of microbes do not respond immediately or consistently to toxic or lethal agents. There are very large differences in the way different kinds of microbes react to the same agent and even within a population of a single type of microbe, there are significant variations in sensitivity. The power of toxic or lethal agents also varies widely. Some of them are intrinsically more active than others and they may be applied differently (more or less concentrated, for longer or shorter periods, more or less hot, etc.). Because of this the outcome of their use can only be predicted in a general way. With very powerful agents the unpredictability is less than with weaker ones and as they get weaker other factors (like the simultaneous presence of dirt) become progressively more important in determining if the process succeeds or fails in its purpose.

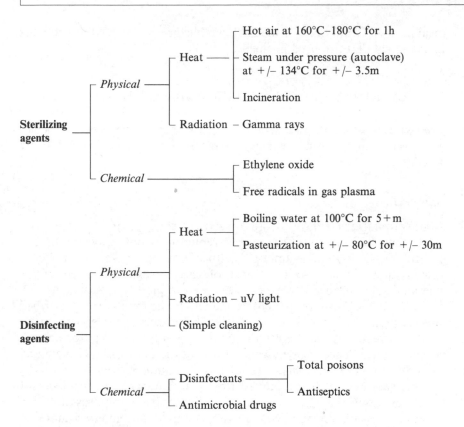

Figure 6.2 A summary of the sterilizing and disinfecting methods commonly used in medical practice. Examples of times are shown in minutes (m) and hours (h).

Figure 6.1 displays the result of applying any killing agent to any population of microbes. This so-called 'killing curve' is divided into three parts. First the killing agent must penetrate to reach the target where it is to act, before it can have any effect. This takes time, represented by part A of the curve. With some agents the time taken may be very short, measured in seconds. With weaker agents it may measure minutes or hours, or if the organism is entirely resistant to the agent, it lasts for ever. Part B of the curve depicts the steady killing (or disablement) of sensitive microbes. The slope of this part of the curve is more or less steep, depending on the rate of killing. This varies with the activity of the agent and the resistance of the microbe. In part C of the curve slightly more resistant microbes take longer to die. With more resistant microbes, large populations and weaker agents, the curve may not reach the baseline, so some microbes survive the process. In different circumstances the time A + B + C can vary from seconds to infinity.

Figure 6.2 summarizes the more important methods employed to disable or kill microbes in medical establishments or by the suppliers of materials that are used in them.

THE RATIONAL USE OF ANTIMICROBIAL DRUGS

The use and abuse of antimicrobial drugs is a proper subject for audit, peer review and, when necessary, control. The options available are described.

Introduction

Antimicrobial drugs have revolutionized the practice of medicine. As a result they have acquired a 'magic' reputation, not only among the general public but also among healthcare workers, many of whom should know better. Because most of these drugs are exceptionally safe and not outrageously expensive, they are greatly misused. At any one time some 30% of patients confined in hospitals are prescribed antimicrobials, either for prophylaxis or for the treatment of infections. Between 1980 and 1991 the number of prescriptions for antimicrobials written each year by general practitioners in the UK increased from 896 to 1453 for every 1000 of their patients (Greenwood, 1995). The increase cannot be accounted for by a change in the incidence of infectious diseases. The overuse of antimicrobials imposes a heavy selective pressure on microbes, so they tend to develop resistance to the drugs most commonly employed (McGowan, 1983).

Controversy has surrounded the use of antimicrobials ever since they were first introduced. Kunin, Tupasi and Craig (1973) studied their use in hospitals. They found that 27% of medical and 29% of surgical patients were prescribed antimicrobials and of the prescriptions, 6% and 50% respectively were written for prophylactic rather than therapeutic indications. Forty-two percent and 62% of them were judged to have been given for inappropriate reasons. 'Inappropriate' may be defined as the lack of a requirement for the use of any antimicrobial or the prescription did not represent the optimum or least expensive choice or it failed to indicate the correct dose, route of administration or the most appropriate duration of therapy. Jogerst and Dippe (1981) reported a survey in which 29% of hospital inpatients received antimicrobials. In this case 40% of the courses were considered inappropriate. The literature contains several other examples of studies that reached similar conclusions.

It is apparent that antimicrobials are often ordered for questionable or inappropriate indications and that prescriptions may indicate inappropriate routes, dosages and durations of therapy. The routine and excessive use of, in particular, broad-spectrum antimicrobials encourages the emergence of multiresistant strains of bacteria. Patients suffer as a result because real infections become more difficult to treat. More expensive and sometimes more toxic drugs have to be used when simpler ones would have sufficed. Infection control teams and committees should encourage the rational use of antimicrobial drugs. Their target ought to be an 'antimicrobial policy' tailored to the needs of each hospital, group of hospitals or community medical service. To be effective these must be based on a clear understanding of local practice in the choice and use of these drugs. This knowledge is acquired by audit (p.66). When patterns of prescribing are known, a policy can be devised to meet local conditions. The methods available for audit and the options for control are discussed below. Antimicrobials are widely employed in the prophylaxis of surgical wound infections. Experience suggests that they are often used for this purpose in ways that are less than optimal, so a description of the scientific background to rational antimicrobial prophylaxis is included here.

The audit of antimicrobial use

Whatever method or combination of methods is chosen, an audit should be carried out under the authority of a drugs and therapeutics committee, an antimicrobial usage subcommittee or their equivalent. Members of the committee should include a representative from the pharmacy service, senior clinicians from all the major clinical departments, a microbiologist and, if possible, an infectious disease physician. This is necessary to secure the co-operation of all of those who might otherwise obstruct the process. Peer pressure is useful when senior individuals who wish to preserve the *status quo* decline to submit to audit or fail to comply with the policy decisions that arise from it.

Kunin (1981) defined a number of methods appropriate to the audit of antimicrobial usage. These may be summarized as follows.

- Gross audit by the examination of pharmacy records.
- An elaboration of gross audit to identify and compare the use of antimicrobials by various clinical services or by individual clinicians.
- The audit of specific situations such as the use of antimicrobials for prophylaxis, in the treatment of individual infections or for any other useful permutation of case review.
- Full audits by incidence or prevalence techniques or as imposed by external bodies for accreditation or the determination of proficiency.

These options vary widely in the amount of work involved and their cost. The organizing committee must select the method, or perhaps different methods applied sequentially, most suitable to their circumstances.

When the audit is complete the committee meets to review the results, to decide what changes are desirable (and attainable) and to formulate these into objectives. The output is likely to be an **antimicrobial policy** (see below). This is published both to inform those who need to be involved and to seek compliance with the committee's conclusions. The committee then plans further audits to monitor the outcome and perhaps to investigate anomalies or other points of interest thrown up by the preliminary audit.

The control of antimicrobial use

In his 1981 paper Kunin identifies a number of methods that might be applied in an attempt to control the use of antimicrobial drugs. The methods proposed are not mutually exclusive and some of them might, with advantage, be used simultaneously. As they formulate a policy the committee will, once more, make decisions appropriate to their circumstances. The options before them may be summarized as follows.

- Education.
- The production of a hospital formulary that restricts the availability of drugs to an agreed list, with a closely controlled let-out clause for exceptional situations.
- To subdivide the agreed list so that new, toxic, expensive or potentially undesirable drugs can only be prescribed by more senior individuals.
- The application of rules that limit the duration of therapy or prophylaxis allowed by a single prescription to a stated maximum duration, only to be extended by the issue of a new prescription.
- The continuous monitoring of prescriptions by clinical pharmacists, with feedback to individual clinicians so they can compare their practice with that of their peers, presented anonymously.
- Restriction of reporting of antimicrobial sensitivity results by the microbiology laboratory to conform with an agreed list of drugs and indications for their use, according to the microbe concerned and the site of the infection.
- To exclude representatives of drug manufacturers or restrict their access to, in particular, junior members of the clinical staff.

The prophylactic use of antimicrobial drugs

Practice in the use of antimicrobial drugs for the prophylaxis of surgical sepsis has nearly turned full circle. Shortly after penicillin became generally available it was grossly abused and staphylococci resistant to it emerged and spread. The result was a notorious pandemic of surgical and other sepsis (Chapter 2). Prophylaxis became a contentious subject and several authorities advised moderation. Between 1945 and 1965 the application of prophylaxis has been characterized as 'peculiarly equivocal (so that) most surgeons concluded that antibiotics had little if any place in the prevention of clinical wound infection' (Polk, 1977). In the special case of contaminated and dirty abdominal surgery, however, their use continued to be explored. In 1969 the value of prophylaxis in this form of surgery was proved by a properly controlled clinical trial. Other trials followed and finally a survey of 26 of them concluded that the benefit of prophylaxis for this purpose was so clearcut that further trials were not justified (Baum *et al.*, 1981). Controlled trials involving other parts of the body then began to restore legitimacy to the use of prophylaxis in clean operations, particularly where surgery is extensive or prostheses are involved. The possibility that prophylaxis might be applied more widely in clean surgery was also discussed (Hopkins, 1991).

It is surprising that so much confusion has surrounded the way antimicrobials are used in the prophylaxis of surgical sepsis. The theoretical basis for its precise application was established about 40 years ago (Miles, Miles and Burke, 1957; Burke, 1961). Prophylaxis is increasingly less effective if given more than an hour after the beginning of surgery and is completely ineffective if the delay is greater than four hours. A sufficient level of an antimicrobial active against as many as possible of the expected pathogens should be present in the wound at the moment of the surgical incision, to be maintained throughout the operation. To avoid unnecessary disturbance of the normal flora the first dose is given as late as is consistent with achieving this end, characteristically immediately preoperatively. A second dose may be required in the course of a lengthy operation (Goldmann *et al.*, 1977). To avoid toxicity due to the drug and to preserve as much as possible of the normal flora, prophylaxis should not be extended beyond the minimum required.

Practice in this respect has varied from a single preoperative dose to multiple doses spread over from 24 hours to five or more days, though the emphasis has been on administration for 24–48 hours. In the case of pelvic surgery the more prolonged use of antimicrobial prophylaxis may represent therapy for an underlying subclinical pelvic sepsis, otherwise differences of opinion may turn on the extent to which a given operation is subject to delayed postoperative infections that are transmitted through the bloodstream. For example, some patients who have undergone total hip replacement are catheterized. The urinary tract infections that are the result have been documented as the sources of such haematogenous infections. Experimental evidence supports the view that extended prophylaxis might delay the onset of urinary tract infection in catheterized patients until after the vascular epithelium of the hip joint has been restored (Howe, 1969).

A more recent clinical trial has helped to validate the theoretical and experimental data described. Classen *et al.* (1992) report the outcome of 2847 clean or clean-contaminated surgical operations performed under the cover of antimicrobial prophylaxis administered at different times in relation to the surgical incision. They record the incidence of postoperative infections among the patients involved (Table 6.1). Statistically significant differences emerged between groups of patients defined by the timing of their prophylaxis.

Table 6.1 The relationship between postoperative sepsis and the time at which antimicrobial prophylaxis was administered in 2847 cases of clean and clean-contaminated surgery (after Classen *et al.*, 1992)

Timing of prophylaxis	No. of operations	No. (%) infected
Pre-incision		
24–2 hours	369	14 (3.8)
2–0 hours	1708	10 (0.6)
Postincision		
0–3 hours	282	4 (1.4)
3–24 hours	488	16 (3.3)

Prophylaxis was most effective when given less than two hours prior to the surgical incision and was significantly less so when the interval between incision and prophylactic administration lengthened, both pre- and postoperatively. If the background rate of sepsis is estimated as the mean of the two worst sepsis rates combined (Table 6.1, a total of 30 infected wounds after 857 operations; 3.5%), then the reduction secured by the most effective use of prophylaxis (ten infected wounds after 1708 operations; 0.6%) is an impressive 83% (2.9/3.5).

The choice of drug used for prophylaxis varies with the range and antimicrobial sensitivities of the bacterial pathogens most likely to be encountered. The predominant determinants are, for the range of pathogens, the anatomical site involved and for their sensitivity patterns, the way antimicrobials are used in the hospital and community concerned. At present a cephalosporin figures in many regimes, with metronidazole added if anaerobes are a threat. New regimes become necessary as patterns of antimicrobial resistance alter and new agents become available or if there is a change in the pathogens responsible for the sepsis it is desired to avoid.

DISINFECTION AND STERILIZATION

The differences between disinfection and sterilization are described, as applied to healthcare, and the principal methods used to achieve them are discussed.

Introduction

The more important factors that determine the outcome when toxic or lethal agents are applied to populations of microbes are the power of the killing agent and the resistance of the microbes. The variability of

Table 6.2 An indication of the sensitivity of different kinds of microbes to killing (disabling) by moist heat

Easy (60°C–80°C)	Intermediate (80°C–100°C)	Difficult (longer at 100°C or >100°C)
Most vegetative bacteria, many viruses	Some vegetative bacteria & spores, some viruses and fungi	Some spores, prions

the response of microbes to heat is illustrated in Table 6.2. A similar table drawn up for toxic chemicals would show the variables as concentration and the length of exposure to them rather than temperature. There would be significant changes in the identity of the microbes in each of the categories. For example, the tubercle bacillus, which falls into the 'easy' category for heat, would move to the 'difficult' one for many of the stronger chemical disinfectants and to a new fourth category ('impossible') for most of the weaker ones.

It is customary to divide the agencies or agents used to kill or disable microbes into those that 'sterilize' and those that 'disinfect'. **Sterilization** is often defined as a process designed to destroy all living things, so microbiological sterility is the total absence of microbes able to reproduce themselves. A sterilized product is therefore no longer infectious, though it may still be microbiologically toxic due to the presence of dead microbes, parts of them or their products ('pyrogens').

This definition of sterilization breaks down because, in practice, absolute and total destruction, ultimately, is unattainable (Kelsey, 1972). It is perfectly possible to design tests so severe that none of the current sterilizing methods would pass them. To overcome this largely theoretical difficulty it has generally been accepted that sterilization is achieved by a process that reliably and reproducibly kills something approaching one million resistant bacterial spores. For steam sterilizers (autoclaves) spores of the heat-resistant *Bacillus stearothermophilus* are used. The unsatisfactory nature of this definition is underlined by the fact that 'sterilization' in an autoclave even to this quite severe standard does not destroy the infectivity of material containing the agents of the spongiform encephalopathies, of which Creutzfeldt–Jakob disease is a human variety and 'mad cow disease' (bovine spongiform encephalopathy) is another example. To inactivate these agents (called prions) they need to be autoclaved for up to six times the normal period (p.173).

A process that kills or inactivates useful numbers of microbes but cannot reliably produce complete sterilization is called **disinfection**. Diagnostic or therapeutic instruments intended for invasive use ideally should be sterile. Because some of them are made of materials that would be damaged or destroyed by the more vigorous process of sterilization, they are disinfected instead. To be effective a disinfecting agent must reduce the number of microbes present on or in an object to a level judged to be harmless in the context concerned and for the purpose intended. What is 'harmless' is difficult to define because a small number of residual microbes, harmless to a healthy adult, might injure an immunocompromised patient under treatment in hospital. Careful attention to the 'purpose intended' is therefore important. Chemical disinfecting agents are called **disinfectants**. If a disinfectant is sufficiently non-toxic (weak) to be used on skin, mucous membranes or exposed tissue, it may be called an **antiseptic**.

In all disinfection and sterilization practice it is important to ensure that objects to be treated are as clean as possible before they are exposed to the process. Cleaning alone removes many of the microbes initially present on or in them, so reduces the chances of failure when they are sterilized and allows satisfactory disinfection to be achieved more easily. For this reason prior cleaning grows in importance as disinfecting methods get weaker. Whenever possible physical methods of sterilization or disinfection are used in preference to chemical ones. This is because they are more easily controlled and more reproducible in their action. They are also usually cheaper and may be safer. They may allow treatment to be completed within a wrapping so the process ends with a dry sterile object that can be stored until it is needed. This is not possible with chemical methods in which toxic liquids are employed because these must be washed away before an object can be used.

Sterilizing methods

Among the sterilizing methods listed in Figure 6.2, those with a physical basis are more powerful and have a wider margin of safety. They may also be more convenient to use than the chemical ones. Chemical methods are employed only when items to be sterilized are made of materials that will not withstand or for some reason cannot be exposed to any of the physical methods. The requirement for sterility (as defined) is imposed by the need to destroy the most resistant, harmful microbial forms. Until recently these were thought to be bacterial spores, particularly those of *Clostridium tetani*, the cause of tetanus. For most other microbes sterilization represents massive overkill, so it provides a wide margin of safety. The growth of interest in prions as causes of human, animal and plant pathology has yet to be reflected in routine sterilizing or disinfecting practice. Far-reaching changes will be required if this ever becomes necessary.

Hot air ovens

Although hot air (or more sophisticated infrared) ovens are comparatively cheap, they are no longer used in many hospitals. This is because disposable items have replaced most of the reusable ones that at one time were sterilized repeatedly by exposure to dry heat at 160°C or more. Another difficulty is that, at these temperatures, wrappings made even of special paper become brittle and fabrics are damaged. To allow storage without recontamination before use, items sterilized in hot air ovens are often packed in custom-made containers, usually made of metal.

A hot air oven resembles a domestic oven fitted with a fan to ensure the even distribution of heat to all parts of the chamber. An accurate thermometer or, better, a recording thermometer regularly checked for its precision is necessary. In use the oven must not be packed tightly so that hot air can penetrate to heat the load as evenly and quickly as possible. Timing of the sterilizing cycle starts when the temperature has reached the desired level: 160°C for one hour is usual for small items that heat up quickly. The use of biological controls (spores) is not necessary as a routine, as the physical parameters (temperature and time) are well established. When a new oven is used for the first time, spore strips containing *B. subtilis* may be included in the load to validate the process. If an unusually large or complex item has to be sterilized by hot air, spore strips may also be used to determine if times in excess of an hour, or of temperatures greater than 160°C are required to allow the heat to penetrate fully. The spores are placed within the packets to be sterilized or in the internal parts of more complex objects. Alternatively, the temperature may be measured with thermocouples or commercially available indicators that incorporate heat-sensitive materials that change their colour or form to show that a certain temperature has been achieved. The latter need to be used with care, as improper storage, for example, may change their sensitivity. Too much reliance should not be placed on these simple indicators.

Steam under pressure – the autoclave

Autoclaves are the mainstay of sterilizing practice in nearly all hospitals. Most reusable items can be sterilized in them, with notable exceptions such as apparatus containing heat-sensitive optical or electronic components (endoscopes or pacemakers) or things made of the kinds of plastics that soften at lower temperatures. An autoclave is significantly more expensive and potentially more dangerous than a hot air oven. Modern fully automatic autoclaves are complex machines that require regular maintenance by expert engineers. They will not work properly, and they may break down, if they are not supplied with steam or water of high quality.

Simple physical principles govern the operation of autoclaves. Those who superintend their use need to understand them if they are to avoid elementary mistakes. The principles involved can most conveniently be introduced by the observation that a hot air oven takes one hour to produce sterility at 160°C, while an autoclave does the same thing in just 3.5 minutes at 134°C, or rather longer at still lower temperatures. The difference is explained as follows.

- Saturated steam condenses on anything cooler than itself, instantly releasing its latent heat of vaporization (540 calories for each gram) so the temperature of the surface concerned rises very rapidly. (Compare putting your hand into an oven at, say, 200°C with the effect on your skin of steam from the spout of a boiling kettle at half that temperature.)
- As steam condenses to water it contracts to a tiny fraction of its former volume. (At 100°C and normal atmospheric pressure 1670 ml of steam turn into 1 ml of water.) The partial vacuum created by this contraction draws more steam to the cooler surface and the process continues until the object to be sterilized has reached the temperature of the steam. Condensation then ceases.
- The temperature of steam rises as its pressure is increased. At normal atmospheric pressure steam is at 100°C, while at twice and three times this pressure the temperatures are 121°C and 134°C respectively (Figure 6.3).

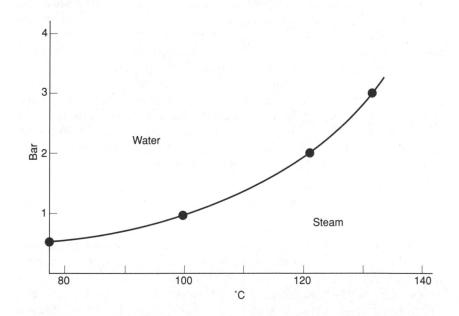

Figure 6.3 The relationship between the temperature at which water boils and the atmospheric pressure. The solid line is the 'phase boundary'. As this line is crossed from upper left to lower right, water turns to steam. This change of state (phase) requires the input of heat energy equal to 540 calories for each gram converted. The temperature does not rise as this happens and this store of hidden energy is called 'latent heat'. When steam condenses back to water the latent heat is released. This is what makes autoclaves work. (Atmospheric pressure, one atmosphere, is equal to one bar, 760 mm of mercury, 760 torr or 106 dynes cm^{-2}. These are approximately equivalent to 29.9 inches of mercury or nearly 15 lbs in^{-2}.)

It follows from the first two points that steam can only sterilize surfaces it is able to reach. Objects with closed inner compartments or that are packed into wrappings or boxes through which steam cannot penetrate are in effect being exposed as if they were in a hot air oven. Because of the lower temperatures and shorter times used in autoclaves, sterilization may not be achieved. Steam is much more effective if most of the air is removed from the autoclave chamber before sterilization commences. Autoclaves operate in various ways to achieve this. The simplest method is by 'downward displacement'. Steam is lighter than air so it collects at the top of the autoclave chamber. As more of it enters, air is progressively displaced downwards, to be discharged through a vent at its bottom. This is a slow and possibly inefficient process (Box 6.1). More efficient autoclaves are equipped with pumps that remove the air. Air removal is always important, but is particularly so in autoclaves designed to sterilize fabrics and wrapped objects, the so-called 'porous load' sterilizers. At the end of a sterilizing cycle a partial vacuum develops inside the chamber as the temperature (and so the pressure) falls. Air is admitted before the door can be opened. This has passed through a filter to remove any microbes that might otherwise compromise the sterility of the load.

Hospital autoclaves operate at two to three times normal atmospheric pressure. The walls of autoclave chambers are made of thick metal to contain this pressure. To avoid accidents this 'pressure vessel' should (and in many countries must by law) be tested regularly to ensure the absence of cracks that might lead to an explosion. During the process of sterilization the metal walls of the chamber are also exposed to steam which condenses on the walls until their temperature rises to equal that of the steam. The water produced wets the contents of the autoclave. Although this is of little consequence for unwrapped metal items, it is unacceptable for anything that is wrapped in materials through which steam must and so water can, pass. When a

---

Box 6.1 A sterilizer that didn't

In April 1971, in a factory in England, about 4000 500 ml bottles of 5% dextrose solution were loaded into a number of autoclaves, each holding about 600 bottles. One of the autoclaves was faulty, and not all the air was removed from the chamber before sterilization began. When the steam entered the residual air collected at the bottom of the chamber, where it surrounded the lowest shelf. The bottles that stood on this shelf were not sterilized. Bottles for quality control, including culture, were taken from the top shelf, so the fault was not detected.

Early in 1972 the bottles reached a hospital in south west England. In the interval the few bacteria left alive in the unsterilized bottles had multiplied into many millions, though on superficial examination the fluid in them appeared normal. Five surgical patients were infused with the fluid, and four of them died. The fluid was extremely toxic, and death was very rapid. An intense search revealed the cause and 54 more contaminated bottles were identified before their use could turn a tragedy into a wholesale disaster.

package is removed from an autoclave a wet wrapping provides a liquid channel through which bacteria can float or even swim to penetrate the sterile interior.

To avoid this, most large autoclave chambers are surrounded by hollow jackets filled with steam. If these are operated correctly the walls of the chamber are heated to just below the sterilizing temperature of the autoclave. This substantially reduces the amount of condensation that develops. The jacket must not heat the chamber above the sterilizing temperature or the steam entering it will be superheated (Figure 6.3). When this happens the autoclave is turned into the equivalent of a much less efficient hot air oven. At the end of a cycle the jacket also helps to dry out the small amount of moisture that condenses in the load as it is sterilized. Of course, the door of an autoclave cannot easily be fitted with a steam jacket, so condensation takes place on the inside of it. The twin doors of a double-ended autoclave not only double this disadvantage, but they also provide extra joints through which air may leak. Double-ended autoclaves are no longer recommended for general use.

Although most of the process is automated, autoclaves pass through certain stages to complete a sterilizing cycle. First, the autoclave is loaded (not too tightly) and the door is closed. Next, much of the air is removed. This is done by various methods, depending on the type and design of the machine. This is followed by the sterilizing phase, during which steam is admitted to raise the load to the predetermined temperature, at which it is held for the required time. Next the steam is removed and air is admitted to return the chamber to atmospheric pressure. Finally the door is opened and the load removed.

Many autoclaves are fitted with controls that allow different sterilizing cycles to be employed. These may be used to sterilize delicate items at lower temperatures for longer times or to process bottles containing liquids. In the latter case the autoclave door must not be opened until the liquid has had time to cool. Bottles that contain liquids at above 100°C may explode when the pressure is reduced. Small, cheaper autoclaves without steam jackets in which steam is generated by boiling water inside the chamber (downward displacement machines) are useful in some settings (outpatient departments, small medical or dental clinics and laboratories, for example). They are suitable for sterilizing unwrapped instruments and utensils required for immediate use or for the sterilization of dangerously contaminated items before they are disposed of. Special autoclaves are employed in pharmacies where significant amounts of liquids are sterilized, particularly when these are packed in plastic containers. Some operating departments have installed autoclaves that operate at higher temperatures to allow very short sterilizing cycles. These 'flash' or 'dropped instrument' sterilizers are sometimes used for the routine processing of items. If this is done the cycles used should be validated in the same way as for any other autoclave.

The twin parameters of time and temperature are the prime physical measures of the sterilizing ability of an autoclave. Temperature and pressure are connected (Figure 6.3) so it is usual to measure both, preferably with an instrument that makes a permanent record. However, the quality of the steam admitted to the autoclave and the amount of residual air in the chamber during sterilization are factors that can cause a failure of sterilization. This may happen despite apparently adequate exposure at a proper pressure and the temperature record may mislead. For practical reasons the temperature inside an autoclave is usually measured at a point other than where it really matters, which is, of course, inside the load being sterilized.

To be at its most efficient, steam must be at what is called the 'phase boundary'. Here it is just saturated, neither containing water droplets (wet or supersaturated steam) nor being superheated (dry steam) (Figure 6.3). This condition is difficult to measure directly, so it is left for designers and engineers to arrange. A more common cause of failure is the presence of too much residual air in the chamber. This may happen because the air-removing system is faulty or, more often, because of a leak. Leaks are always important, but in a porous load autoclave even an invisible hole in a door seal or at a joint between pipes can admit a significant amount of air during any vacuum phases in the sterilizing cycle.

Testing autoclaves

There are a number of different ways in which autoclaves may be shown to have worked correctly. Reliance may be placed on physical parameters (time, temperature, pressure) or on biological indicators (bacterial spores) or on physical or chemical devices that change their colour or form on exposure to a certain temperature for a certain time, or on any combination of these. The choice of method is the subject of a debate that has sometimes been as hot as the steam being tested!

Physical measurements are precise and objective and the results are available immediately, so a load can be passed as sterile at once. They suffer from the disadvantage that the measurements are recorded by instruments that must be calibrated regularly to ensure accuracy and as noted, steam quality and air removal are important though less easily measured components of success. When physical parameters are used to monitor an autoclave the routine measurement of temperature and pressure must be backed up by a programme of extra physical tests that include a search for leaks and the use of thermocouples to make direct measurements of the temperature achieved at various points in the chamber. These tests are done on a regular basis by trained engineers who use special instruments. A major advantage of this approach is that the measurements can be made with such sensitivity that small progressive deteriorations in performance can be detected before a fault affects sterilization. This allows corrective maintenance to be a planned, rather than an emergency activity.

Sudden catastrophic failures still happen, of course, but these are covered by the daily performance of a Bowie–Dick test.

The **Bowie–Dick test** is a test for air removal and steam penetration, so is only applicable to porous load autoclaves. The test should be performed as the first run of each autoclave every working day. The pack used is placed at about the centre of an otherwise empty autoclave. The traditional pack is made up of not more than 36 clean, dry, preferably huckaback hand-towels. Each is folded into eight thicknesses and they are stacked until an approximate cube is formed 11–12 inches (280–305 mm) high. A piece of paper bearing an 'X' of 3M brand autoclave tape is placed between the towels at the centre of the stack. When the autoclave cycle is complete the stack is removed and the tape examined. If the tape has changed colour evenly right across both arms of the 'X' the autoclave has passed the test. A failure is indicated by the colour change being less intense at the centre than at the outside end of each arm of the tape. This is due to residual air that has been compressed into a bubble at the centre of the pack, where the two arms of the 'X' cross over. The bubble excludes steam and so prevents sterilization. If an autoclave fails the Bowie–Dick test it must not be used until it has been repaired. Commercial equivalents of the home-made Bowie–Dick test pack are available.

Biological tests challenge the autoclave to kill something approaching a million spores of *B. stearothermophilus*. One or more paper strips impregnated with the spores enclosed in glassine envelopes or other steam-permeable sachets are included inside typical packs to be autoclaved. On completion of the cycle the strips are extracted and cultured microbiologically to detect any spores that may have survived. Spore strips or kits that contain the spores and the necessary culture medium so that the test can be carried out within the department housing the autoclave are available commercially. As with any biological test, a positive control (a strip from the same batch that has not been autoclaved) must be included with each set of cultures to show that the spores were viable at the outset. The test takes at least 48 hours and perhaps seven days to complete. For safety the contents of the autoclave may be stored until the result indicates that sterilization was successful. If biological tests are adopted as the prime method used, this period of quarantine has cost implications. A failed test requires the autoclave to be taken out of use until the fault has been found and corrected. The test can only give a negative or a positive result, so no advance warning of impending autoclave failure is provided. Biological tests are poorly reproducible, though because of the degree of overkill in an autoclave this may not matter. When porous load sterilizers are monitored biologically it is customary to use the daily Bowie–Dick test as well.

A number of commercial devices are available that may be applied to the outside of a pack or put into it during sterilization. These change their colour or some other aspect of their appearance when exposed for a certain time at a particular temperature. Once they have been validated by an independent authority, they must be used with strict compliance to their manufacturers' instructions, particularly with regard to storage conditions and shelf life. They are generally less satisfactory than the other control methods described, though indicator 'autoclave tape' on the outside of a pack provides visual reassurance that it has, in fact, passed through an autoclave at all!

Arguments about which method to use for the control of autoclaves are usually as sterile as their contents ought to be. The potential for error introduced by human fallibility is likely to be greater than the effect of the differences between them. The choice depends on circumstances. Physical methods require that qualified engineers are available near at hand to perform the necessary regular tests. To make this cost effective, each engineer needs to cover a large number of autoclaves. The other methods are cheaper and more appropriate in parts of the world where small hospitals are separated by large distances. In this situation it has been customary to use autoclaves that operate at lower pressures and temperatures (121°C) in a longer cycle. Provided those who use them are acquainted with the weaknesses of the test method they adopt, all are more or less satisfactory.

The incinerator

Fire simultaneously sterilizes and consumes what is burnt. On the face of it incineration provides the ideal way of dealing with hospital waste. In one operation waste is reduced to a negligible volume and is freed of obnoxious characteristics and any infectious potential.

Unfortunately this is not the whole story (p.203). Apart from the long-term ecological damage they may do, incinerators have several more immediate disadvantages. To achieve destruction when the material sent for incineration contains much liquid (placentas, incontinence pads, disposable suction bottles, etc.), the process will consume significant amounts of oil or gas. Most hospital waste includes a (perhaps large) proportion of plastic. When the amount of certain types of plastic rises too high dense black smoke is produced that is unacceptable to neighbours and clean air inspectors alike. An invisible but more damaging byproduct of burning plastic is hydrochloric acid which, before it passes into the atmosphere to cause pollution, attacks the incinerator itself and so shortens its life. In the case of simple incinerators there is concern that small particles, perhaps carrying microbes, might be swept up the chimney by the flue gases before they are completely burnt. The distant possibility that this might do harm, together with the probability that certain toxic gases are emitted, has imposed expensive modifications on the design of hospital incinerators. In some cases, where they are still used, the difficulties described have limited the volume of material that can be processed and elsewhere they have had to be closed down. In some states ecological concerns have led to restrictive or prohibitive regulation or legislation that has so complicated the local destruction of hospital waste that it has sometimes been smuggled out for disposal in less discriminating countries.

Radiation

Cost, complexity and very stringent safety requirements surround the use of ionizing radiations (usually gamma rays) for sterilization. This has limited the practice almost entirely to large industrial facilities that deal with disposable items such as syringes and needles.

Chemical methods

Although other chemicals (notably formaldehyde – see Disinfecting methods: Heat, below) have been used in sterilizing processes, only ethylene oxide (EO) is commonly employed as a true sterilant for medical purposes. New entrants to this field are so-called 'free radicals'.

Ethylene oxide

EO is a non-corrosive, highly effective sterilizing agent, with good penetration. Unfortunately it is flammable and explosive when mixed with air at certain concentrations. It is also irritant, toxic, mutagenic and carcinogenic. It is colourless and remains odourless at concentrations well above those that are toxic. Personnel may be exposed to danger without being aware of it, unless sensitive automatic detectors are fitted in places where the gas is used. Personal monitors may also be worn. In some countries concern has grown about use of EO and the maximum level to which workers may legally be exposed has progressively been reduced. If an effective substitute were to appear it is likely that this maximum would be reduced still further, to the point at which the use of EO would become uneconomic.

Successful sterilization with EO depends on more variables than apply to steam under pressure. Time, temperature, humidity, pressure and gas concentration are all important. Some machines work at increased pressure and in these cases the EO is mixed with an inert gas such as carbon dioxide or one of the ecologically less friendly chlorofluorocarbons (CFCs). This technique reduces the risk of an explosion, but increases the likelihood of a leak. Other machines work at subatmospheric pressures. Different manufacturers use different sterilizing cycles.

Items sterilized in EO require lengthy aeration to free them of residual gas before they are used, particularly if they are wrapped. To reduce the exposure of staff when unloading machines, initial aeration may be carried out for the first 18 hours or so (overnight) inside the sterilizing chamber itself. Following this, further aeration is required which varies with the material concerned (rubber, silicone, polyvinyl chloride take longer) and the temperature at which the process is carried out. Even at a high temperature (55°C), 12 hours is necessary.

Because many of the physical factors on which the production of sterility depends are difficult to measure, EO machines are monitored biologically. In this case spores of *B. subtilis* are usually employed, otherwise in the manner described for steam sterilizers. As there are no other controls spore strips are exposed in every cycle. One strip suffices in a small sterilizer, though several are needed in larger machines. If items with long narrow lumens are to be sterilized a strip should be placed at the most distal or difficult point. If this is impossible a special test piece is used (Line and Pickerill, 1973) or one may be constructed locally by attaching a narrow tube to the cap of a small bottle in which the spore strip is placed. The spores are usually retrieved for culture after preliminary aeration in the sterilizer chamber is complete. Once more, the need to quarantine packs until a negative spore culture is recorded has cost implications. Many of the items sterilized by EO are expensive so may not be available in such numbers as to allow them to be out of use for the period necessary. Part of this time is, of course, occupied by the aeration phase.

All sterilization processes should be supervised by specially trained personnel, but this is particularly true in the case of EO machines. The number of items for which EO sterilization is an absolute requirement is small and many of the things for which it is more often used can be made microbiologically safe by other methods. EO machines are expensive, potentially dangerous and have high running costs. They should not be installed without careful thought.

Free radicals

Radicals are atoms or ions that have been brought to a chemically highly excited, reactive state. They are incapable of prolonged independent existence. They may be produced in chemical reactions and by the discharge of electricity or the passage of radio or microwaves, through highly rarefied gases (in a near vacuum). The gas may glow as a result. Neon lights and the Aurora Borealis ('northern lights') are manmade and natural examples of this effect. A gas excited in this way is called a **plasma** and the atoms or ions it contains are **free radicals**.

The idea of sterilizing surfaces by the application of free radicals in a gas plasma was patented in 1968. The idea is attractive because the highly reactive radicals rapidly recombine to produce non-reactive and so safe endproducts. This makes the method a possible substitute for EO and a number of workers have evaluated the idea. It was found that more reactive gases (oxygen, hydrogen, hydrogen peroxide vapour) produce more active plasmas than those that are less reactive (argon, nitrogen). In order to sterilize, exceptionally high vacuums are required. There are even more variables in the system than with sterilization by EO and rather sophisticated machinery is necessary. Penetration is poor and special accessories are needed to ensure the plasma reaches into the long thin channels of endoscopes. Tubes with 'dead ends' or complex items will require validation with spore strips to confirm that they have been sterilized. As happens with other chemical disinfectants, the radicals react actively with and are used up by any organic material that is present, so items for sterilization must be scrupulously clean. Commercial gas plasma (free radical) sterilizers have been approved by the US Food and Drugs Administration (FDA) (see Crow and Smith, 1995). In its present state of development, the method has yet to prove itself as a complete substitute for EO.

Disinfecting methods

Disinfecting methods (summarized in Figure 6.2) have a very broad spectrum of efficacy. At one end are the weaker methods with such low general toxicity that they can be used on or in living tissues (antiseptics). At the other are very powerful toxic agents that can only be applied to 'things'. In ideal circumstances the latter may produce sterility but as this cannot be guaranteed and the margin of safety is narrow, they are included with the disinfectants. Although neither sterilization nor disinfection is an instantaneous event, with disinfection the time element is more important and more difficult to monitor. Many sterilizers are fitted with devices that prevent premature opening: indeed, the operator might find it fatally dangerous to do so. Disinfection takes longer, sometimes much longer. The process is often directly accessible to the person who applies it, who may also use the end-product. Natural impatience may cut the time of its application to the point where even disinfection is not achieved. The great importance of thorough cleaning before items are disinfected is stressed once more.

Heat

The application of heat to produce disinfection is usually by the use of hot water or steam. Steam applied in a steamer at atmospheric pressure (about 100°C) is now rarely used outside laboratories. Specially designed subatmospheric low-temperature steam autoclaves (originally operating at 355 mm of mercury and 80°C, Figure 6.3, though now often at 73°C) are available (Alder, Gingell and Mitchell, 1971). These produce a very high order of disinfection in wrapped items that might be damaged by higher temperatures. If formaldehyde is added to the process sterility can be achieved. Technical difficulties have not yet been overcome and low-temperature steam and formaldehyde machines are not widely used. Formaldehyde would be an attractive alternative to EO not least because the human nose can detect it at levels below that at which toxicity is thought to be a problem.

Boiling water is an extremely efficient, high-level disinfecting agent.

It fails to kill only a few resistant spores of kinds rarely found on clean surfaces. It was given a bad press during the campaign that accompanied the transfer of the preparation, disinfection or sterilization of instrument and dressings to operatives in central supply organizations. This task had been performed by nurses in hospital wards and departments. It is unlikely that any clean instrument that had been boiled for not less than five minutes ever caused an infection. The argument that instrument 'boilers' are dangerous to staff who regularly use boiling water for making tea or coffee is not easily sustained. The main problem with boilers was that they were abused by being overfilled by multiple users, so that no-one could tell if a particular item had been in water (that may have ceased to boil) for 30 seconds or three hours. Another problem was how to disinfect the notorious Cheatle forceps with which items were removed from the boiler. When all is said and done, however, it is well to remember that in an emergency properly timed exposure to water that is really boiling is a better, quicker, cheaper, more readily available and more convenient near-sterilizing method than any other.

For items that cannot be boiled, pasteurization may be carried out in water at temperatures between 60°C and 90°C for times varying with temperature from a few seconds to an hour. This is a very effective disinfecting agent: properly pasteurized milk is very safe milk. The process is widely applied in the washer-disinfectors that are used for the cleaning and disinfection of cutlery and crockery, non-disposable bedpans and urinals, anaesthetic apparatus, for cleaning used instruments prior to repackaging and sterilization and in the laundry.

Ultraviolet light

Short-wavelength ultraviolet (uV) light can kill or inactivate microbes. Sunlight is quite a good disinfecting agent. Artificial sources that emit sufficient short-wavelength uV light are effective antimicrobially, provided nothing obstructs the radiation. The light does not penetrate far into liquids and it is stopped by dust or other dirt on the glass of the source lamp or on the item to be disinfected. As a lamp ages the wavelength of the light it emits lengthens, so although the lamp appears to be operating normally, it may be deficient in antimicrobial activity. Special apparatus is required to detect this gradual loss of efficiency. Experiments have been performed in the use of uV radiation to treat the air in occupied spaces, to prevent infections. This approach failed to reduce respiratory infections in schools and has had little influence on surgical sepsis (p.146). Another application has been in the production of near-sterile water for clinical purposes. With a new lamp, clean surfaces and water of good initial quality this works quite well, but deterioration due to the factors mentioned results in a loss of efficiency that will not be noticed unless regular monitoring is carried out.

Disinfectants and antiseptics

Bad smells are very often the byproducts of bacterial putrefaction. For this reason the human nose can detect the presence of some kinds of bacteria and this led to the use of certain disinfectants before microbes were discovered. Some chemicals were found to control smells and because smells were associated with disease (Chapter 2), the same chemicals were used to try to control infections. Chlorine and carbolic acid were employed in this way before the science of microbiology was established. They are still used as hypochlorite and phenolic disinfectants. A number of other chemicals have joined them. It is notable that a direct relationship exists between the degree of animal and human toxicity of a disinfectant and its general ability to kill or inactivate microbes. Very powerful disinfectants are extremely toxic to all living matter, while disinfectants (antiseptics) that can safely be used on skin or mucous membranes are significantly weaker and may only be effective against certain classes or types of microbe. Some hardy bacteria that are able to grow in solutions of weak disinfectants have caused infections in patients on whom supposedly 'sterile' antiseptics had been used (Box 6.2). Weak disinfectants may not even disinfect themselves.

Disinfectants, and particularly the weaker antiseptics, are best supplied sterile, in single-use containers. If this is not possible disinfectant concentrates may be stored in bottles that are reopened occasionally, but once a disinfectant has been diluted it should be used immediately and any residue discarded at once. Bottles of stock disinfectants or antiseptics should never be 'topped up', but fresh clean, dry, preferably sterile bottles used each time. Some disinfectants are inactivated by cork, soap, hard water, plastics or rubber. Products bought from reputable manufacturers will include warnings about this in their literature.

Box 6.2 An antiseptic that wasn't

Among the bacteria most commonly found as contaminants in supposedly sterile medications, members of the genus *Pseudomonas* figure prominently. Pseudomonads are notoriously resistant to the weaker disinfectants, so are found rather commonly in antiseptic solutions that have been prepared or handled incorrectly.

A hospital in the USA used an iodophor preparation for the disinfection of the insertion sites of the indwelling catheters used for continuous ambulatory peritoneal dialysis. In a period of just over a month five patients developed infections, four with peritonitis and one of the skin at the catheter insertion site. In each case cultures revealed the presence of *Pseudomonas aeruginosa*. A search for the source led to the iodophor solution and *Ps. aeruginosa* with the same sensitivity pattern was isolated from previously unopened one-gallon containers of the iodophor. Antiseptics are such weak disinfectants they may not even be able to disinfect themselves!

All disinfectants attack organic matter more or less indiscriminately and are used up in doing so. If microbes are enmeshed in a large amount of dirt, a disinfectant applied to the mixture may be exhausted before all the microbes are killed or indeed before it has penetrated much below the surface. This is why objects to be disinfected should, whenever possible, be cleaned beforehand. When this is not possible more concentrated preparations of some disinfectants can be used, which may be described in the accompanying literature as appropriate for 'dirty' situations.

Because of the wide variations in availability, national regulation and prejudice, only general observations are made about individual disinfectants. Each medical establishment should have its own disinfectant policy, prepared by the infection control team or committee responsible for it. It is an expensive mistake to provide a large range of different disinfectant products to cater for a multitude of individual preferences. It is important to note that many of the most advanced hospitals have abandoned the use of disinfectants in environmental cleaning, replacing them with a simple detergent in water (p.250).

Many hospitals will find that one disinfectant from each of four or five of the general classes described below will suffice, though more than one formulation of a particular product may be required.

- Where it is still thought necessary, a powerful general purpose agent is provided for the disinfection of accidental spills of potentially infected body fluids, etc. This must be active against all microbes. Chlorine as hypochlorite is most commonly used for the purpose. If purchased as a liquid the most concentrated form generally available commercially is 10% available chlorine or 100 000 parts per million (ppm). Hypochlorite may be used for spills at 1% (10 000 ppm), at 0.1% (1000 ppm) for general environmental disinfection and at 0.01% (100 ppm) for the disinfection of teats and other infant feeding equipment. Tablets of sodium dichloroisocyanurate that dissolve in water to produce a hypochlorite solution provide an alternative source of chlorine for disinfection. Although initially more expensive, the tablets are more easily stored, more easily used and are perhaps less wasteful. They are available from a variety of commercial sources. Hypochlorite is a highly effective disinfectant, but may corrode metal and damage plastics, rubber and fabrics. Solutions, particularly when more dilute, quickly lose their potency. Although concentrated hypochlorite may damage metal instruments, it is non-toxic at low concentrations and safe to use for catering equipment. Hypochlorite is widely applied at a concentration of about 5 ppm available chlorine (or less) in drinking water and the water in swimming pools. Note that chlorine has been used in war as a poison gas. It would be difficult to generate significant amounts of the gas from disinfectant products, but they should be treated with respect and not employed in enclosed, poorly ventilated spaces.

- Glutaraldehyde and some other aldehydes, including formaldehyde, are kinder to metal and plastics than hypochlorite. Glutaraldehyde is available in various forms from several commercial sources. Its principal use is to disinfect endoscopes. It is important to pay particular attention to the literature provided by the manufacturers of both the disinfectant and the endoscope and to heed the advice of various professional bodies in the field. Although sometimes described as a sterilant, it is usually applied for too short a time to achieve sterilization. While most bacteria are killed within two minutes, HIV and the hepatitis B virus may take 10–30 minutes, tubercle bacilli 60 minutes and spores 3–10 hours. Glutaraldehyde is toxic, irritant and allergenic. Gloves must be worn when handling it, tanks of it must be covered and rooms where it is used must be well ventilated (in some countries more stringent rules have been introduced). Of course, an instrument soaked in it must be thoroughly rinsed in sterile water before it is used. Peracetic acid (an equilibrium mixture of hydrogen peroxide and acetic acid that react together to produce a third component, peracetic acid itself) is an alternative to glutaraldehyde. Although unpleasantly toxic in its concentrated form, it eventually degrades to oxygen and acetic acid (vinegar) so it is more environmentally friendly.

  Special types of equipment ('**chemical washer-disinfectors**') have been devised and are commercially available for the disinfection of endoscopes with these agents. Fluids are actively pumped through the various endoscopic channels, so the process is quite efficient. These increasingly sophisticated machines do not replace the initial manual wash that is still vitally necessary before chemical disinfection is attempted. This must be accompanied by a careful visual scrutiny for traces of dirt in crevices and the joints of moving parts and, particularly in hard water areas, for the development of a biofilm. If endoscopes are simply soaked in a disinfecting fluid, two points require attention. First, if instruments are wet when immersed in the solution, significant dilution of the disinfectant may result as successive instruments are treated. Second, care must be taken to ensure that the disinfectant and rinsing solutions reach into all parts of the apparatus. In particular, there must be no bubbles in narrow channels. The chances of this are reduced if instruments are immersed vertically in tall tanks rather than being laid flat in trays. It has yet to be decided how best to provide the large volumes of sterile water needed to rinse endoscopes after they have been disinfected.

- If still thought necessary, phenolics may be used for environmental disinfection in situations in which bacterial pathogens are to be attacked and hypochlorite is inappropriate. A large range of 'clear soluble phenolics' is available from commercial sources.

- Alcohol as ethyl alcohol (spirit) or as isopropyl alcohol, diluted in water to 70%, is a very effective disinfectant. It is particularly useful because it evaporates and so does not have to be washed off after disinfection and surfaces are left dry. Alcohol does not kill spores, but is active against all other microbes. It can be used on intact skin where it is the best disinfectant for general use, but it must be allowed to evaporate because prolonged contact with alcohol in concentrated liquid form can cause a chemical burn. It is used on its own or may be combined with another antiseptic to prolong its effect. It can be distributed in liquid form or impregnated into paper tissues packed individually or multiply. These are useful when disinfecting skin or other surfaces. Alcohol is flammable.

- There are a number of skin and tissue antiseptics, of which iodophors (such as povidone iodine) and chlorhexidine are the most widely used. Either may be employed in aqueous solution for use on mucous membranes or exposed tissues. Combined with a detergent, they may be used as antiseptic surgical scrubs or for cleaning dirty wounds. Formulated with alcohol they are useful as skin disinfectants that combine a powerful immediate effect and a residual activity. They may be made up as ointments. Povidone iodine has a broad antimicrobial spectrum, though it acts rather slowly, particularly against spores. Chlorhexidine is more active against Gram-positive than Gram-negative bacteria, but has no activity against tubercle bacilli or spores. It is effective against fungi and some viruses, including HIV. Another antiseptic, hexachlorophane, is even less active against Gram-negative organisms but it may have a particular role in the control of staphylococci (Box 6.3). Triclosan is a somewhat less effective alternative to hexachlorophane.

Summary

Recent decades have seen a highly significant reduction in the application of chemical disinfectants to the environment in hospitals and

---

Box 6.3 Another antiseptic that wasn't

In the late 1960s, patients in a surgical ward in a military hospital began to suffer from postoperative wound infections with an unusual Gram-negative pathogen, *Serratia marcescens*. A number of wounds broke down and the surgical and nursing staff became alarmed. An initial search failed to reveal the cause and the imposition of strict hygiene made no difference. Finally an antibacterial surgical scrub, previously only used in the operating rooms, was introduced into the ward for handwashing. The epidemic not only continued, but worsened. Suspicion fell on the scrub, a liquid preparation that contained hexachlorophane. Culture yielded a heavy growth of *S. marcescens*, indistinguishable from that found in patients' wounds. Further investigation showed that the organism was growing in the one-gallon containers in which the product was received from the manufacturers. The scrub was withdrawn and the incident came to an end.

   The outbreak originated in the operating rooms, where the staff were using a heavily contaminated surgical scrub. When this was introduced into the ward it made matters worse. Because the material was labelled 'antiseptic' its disinfectant properties were taken for granted!

other public buildings. Epidemics of infection have not resulted, money has been saved and the environment spared contamination with unfriendly chemicals. In healthcare, current 'best' practice is to restrict disinfection almost entirely to the skin and mucous membranes, to flexible endoscopes, for specific purposes in laboratories and (with a growing lack of conviction) to spills of body fluids. Additional valid uses are identified in food preparation, so hospital kitchens are included in the diminishing list of places where the use of disinfectants has not yet been actively discouraged.

Despite these restrictions, 'new' disinfectants appear regularly. Hydrogen peroxide, chlorine dioxide, ozone and peracetic acid have all emerged from use in other areas to enter the healthcare field. These and a few others, including a peroxygen compound, are more or less powerful oxidizing agents that, under ideal conditions, may produce sterility in the sense defined above. Their margin of safety is, however, narrow compared with that provided by an autoclave. They are all inactivated by the organic materials commonly found on objects that need to be 'sterilized' and their penetration is poor. It is safer to think of these chemicals as high-level disinfectants rather than as sterilants. As oxidizing agents, as well as the microbes that are their primary targets, they may also attack some other materials such as fabrics, plastics, rubber, metal, etc. It is necessary to check on the possibility of chemical damage before expensive items are exposed to them. When confronted by an enthusiastic salesperson with a 'new' disinfectant, a healthy scepticism is the best initial response!

REFERENCES

Alder, V. G., Gingell, J. C. and Mitchell, J. P. (1971) Disinfection of cystoscopes by subatmospheric steam and steam and formaldehyde at 80°C. *British Medical Journal*, **ii**, 677–80.

Baum, M. L., Anish, D. S., Chalmers, T. C. *et al.* (1981) A survey of clinical trials of antibiotic prophylaxis in colon surgery: evidence against further use of no-treatment controls. *New England Journal of Medicine*, **305**, 795–9.

Burke, J. F. (1961) The effective period of preventive antibiotic action in experimental incisions and dermal lesions. *Surgery*, **50**, 161–8.

Classen, D. C., Evans, R. S., Pestotnik, S. L. *et al.* (1992) The timing of prophylactic administration of antibiotics and the risk of surgical-wound infection. *New England Journal of Medicine*, **326**, 281–6.

Crow, S. and Smith, J. H. (1995) Gas plasma sterilization – application of space-age technology. *Infection Control and Hospital Epidemiology*, **16**, 483–7.

Goldmann, D. A., Hopkins, C. C., Karchmer, A. W. *et al.* (1977) Cephalothin prophylaxis in cardiac valve surgery. *Journal of Thoracic and Cardiovascular Surgery*, **73**, 470–9.

Greenwood, D. (1995) Sixty years on: antimicrobial drug resistance comes of age. *Lancet*, **346** (Supplement), 1.

Hopkins, C. C. (1991) Antibiotic prophylaxis in clean surgery, peripheral vascular surgery, noncardiac thoracic surgery, herniorrhaphy and mastectomy. *Reviews of Infectious Diseases*, **13** (Supplement 10), 869–73.

Howe, C. W. (1969) Experimental wound sepsis from transient *Escherichia coli* bacteraemia. *Surgery*; **66**, 570–4.

Jogerst, G. J. and Dippe, S. E. (1981) Antibiotic use among specialities in a community hospital. *Journal of the American Medical Association*, **245**, 842–6.

Kelsey, J. C. (1972) The myth of surgical sterility. *Lancet*, **ii**, 1301–3.

Kunin, C. M. (1981) Evaluation of antibiotic usage: a comprehensive look at alternative approaches. *Reviews of Infectious Diseases*, **3**, 745–53.

Kunin, C. M., Tupasi, T. and Craig, W. A. (1973) Use of antibiotics: a brief exposition of the problems and some tentative solutions. *Annals of Internal Medicine*, **79**, 555–60.

Line, S. J. and Pickerill, J. K. (1973) Testing a steam formaldehyde sterilizer for gas penetration efficiency. *Journal of Clinical Pathology*, **26**, 716–20.

McGowan, J. E. Jr (1983) Antimicrobial resistance in hospital organisms, and its relation to antibiotic use. *Reviews of Infectious Diseases*, **5**, 1033–48.

Miles, A. A., Miles, E. M. and Burke, J. (1957) The value and duration of defence reactions of the skin to the primary lodgement of bacteria. *British Journal of Experimental Pathology*, **38**, 79–96.

Polk, H. C. (1977) The prophylaxis of infection following operative procedures, in *Hospital-acquired Infections in Surgery*, (eds H. C. Polk and H. H. Stone), University Park Press, Baltimore, pp. 99–107.

FURTHER READING

Ayliffe, G. A. J., Coates, D. and Hoffman, P. N. (1993) *Chemical Disinfection in Hospitals*, Public Health Laboratory Service, London.

Coates, D. and Hutchinson, D. N. (1994) How to produce a hospital disinfection policy. *Journal of Hospital Infection*, **26**, 57–68.

7 Characteristics and control of specific iatrogenic infections

URINARY TRACT INFECTIONS

Iatrogenic urinary tract infections: basic data, pathogenesis, diagnosis, treatment and control.

Basic data

Most surveys of iatrogenic infections have shown that the anatomical site most often involved is the urinary tract. In the prevalence surveys listed in Tables 3.1 and 3.2 (pp.29 and 30) urinary tract infections (UTIs) accounted for between 23% and 45% of the IaIs that were recorded. When these surveys are arranged in date order they show a decline in the prevalence of iatrogenic urinary tract infections from an average of 43 per thousand hospital patients from 1961 to the end of 1980 (39% of IaIs) to 25 per thousand between 1981 and 1995 (31% of IaIs). This almost certainly reflects the greater and more careful use of closed drainage systems in association with urinary catheterization.

In what follows, more numerical and other data are provided to illustrate the role of UTI as the most common cause of IaI. These have been taken from a variety of sources, so the picture presented is a composite one (unless otherwise noted, see reviews by Stickler, 1990; Stamm, 1991, 1992; Anon, 1991; Garibaldi, 1992).

Most iatrogenic UTIs appear to be endemic, as befits what has been perceived as an endogenous infection. Detailed typing of the microbes responsible, however, suggests that in a significant number of cases their causes may be the autogenous variety of self-infection, so they lie in the grey area between self- and cross-infections (p.48). There are strong associations between these infections and urethral instrumentation, particularly catheterization. Urethral catheters are thought to be the cause of about 80% of the UTIs acquired in hospitals and other forms of urethral instrumentation probably account for another 10%. In a prevalence study 21% of patients who were catheterized had developed UTI compared with only 3% of those who were not. The same study found that, at any one time overall, 10% of patients in acute general hospitals were catheterized and that the proportion was higher among the elderly (Meers *et al.*, 1981). The duration of catheterization varies, but another study showed that half of them had been removed by the fourth and 80% by the tenth day (Mulhall, Chapman and Crow, 1988). Catheters are most commonly used in the postoperative care of surgical patients, in cases of bladder neck obstruction and incontinence or for the measurement of urine output.

With closed drainage (Figure 7.1) and average-to-good conditions of care, just under 10% of uninfected patients are newly infected each day catheterization persists, to reach a total of 50–60% by the tenth day. Rates of acquisition above this require urgent investigation to find out what has gone wrong. Lower rates do not call for congratula-

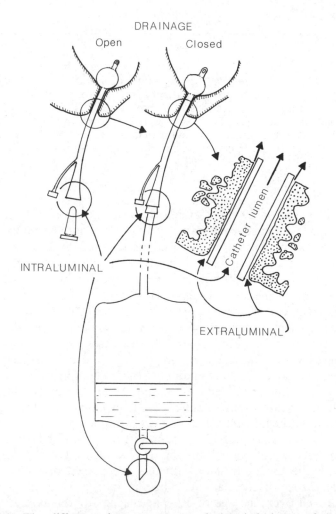

Figure 7.1 The difference between open and closed drainage of the urinary bladder and the routes by which microbes may gain access to it.

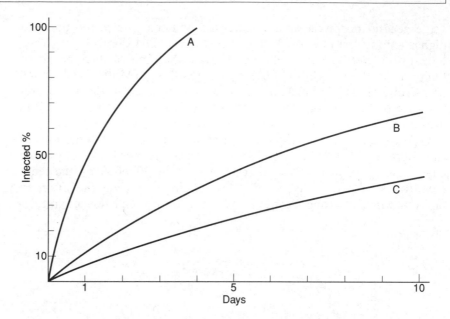

Figure 7.2 Curves representing the acquisition of urinary tract infections in a cohort of mixed sex, catheterized patients: A, with open drainage; B, with closed drainage and new infections at a rate of 10% of the population each day; C, the same, at 5% a day. For use as performance indicators in short term catheterization, rates of acquisition that lie between A and B are unacceptable, between B and C fair but might do better, and below C, increasingly good.

tion until figures of 20–30% at ten days are achieved (under 5% of patients newly infected each day). With open drainage almost 100% of catheterized patients are colonized or infected by the fourth day. With particularly well-managed closed drainage this figure is not reached until about the 30th day. This information may be used to set preliminary performance indicators as measures of success or failure in the management of urinary catheters (Figure 7.2).

Septicaemia (to be distinguished from bacteraemia; see Glossary) is said to develop in between 1% and 3% of cases of UTI. Septicaemia is associated with a mortality of up to 30% (more in severely debilitated patients) so is an outstandingly serious complication. The high frequency of catheter-related UTI highlights the urinary catheter as the single most important cause of death in cases of IaI. In practice most individual healthcare workers meet examples of this complication infrequently and may fail to recognize them when they do, so this important fact is not widely known. The result is that the risks associated with the use of urinary catheters are played down or ignored (Box 7.1). Most of the patients who die in this way are over, and many are well over, the age of 50. The reason for this is that patients over this age are catheterized more frequently and are also more likely to be debilitated and so predisposed to infection.

Box 7.1 Urinary catheters and septicaemia

If the incidence of iatrogenic infection is taken as 5.7%, and the median figure for the proportion of it due to UTI is 36% (Table 3.2) then of each 10 000 patients admitted to a hospital, about 200 will develop iatrogenic UTIs (or if the rate of catheterization is 10%, 200 of every 1000 patients catheterized). If 2% of those infected develop septicaemia with a one-third mortality, there will be four cases of septicaemia, and one death. This may be a conservative estimate (see Chapter 5, p.84). The minimum overall position may be summarized as follows.

- Patients admitted   10 000
- Catheterized   1000
- Infected   200
- Septicaemic   4
- Dead   1

If the rate of infection could be halved, think what would be saved!

Pathogenesis

With the exception of the distal part of the urethra, the healthy urinary tract is sterile. The distal part of the urethra is colonized by members of the normal, usually aerobic, flora of the perineum, which includes Gram-negative bacteria from the gut. Periodic voiding and the constant flow of urine into the bladder restricts the backward spread of colonization into the rest of the urinary tract and this is aided by the secretion of bactericidal substances from the prostate in the male and the periurethral glands in the female.

Most bacteria grow well in urine. Lister used urine as his culture medium when he repeated Pasteur's experiment that disproved the spontaneous generation of life (p.8). Urine is inanimate so cannot itself be infected, but a sufficiently heavy colonization of urine while still within the urinary tract (a condition described as **bacteriuria**) eventually leads to inflammation of the tissues that line it and so to UTI. The urinary tract is most commonly invaded from below through the urethra (an ascending infection) or less commonly from the blood via the kidneys (a descending infection). Rarely, infections may extend from neighbouring structures. The transient, symptomless presence of bacteria in urine is common, more so among females as a consequence of their shorter urethras.

Factors that predispose to UTI are any structural or functional abnormality of the tract that can provide a foothold for the multiplication of microbes in living tissue, to convert asymptomatic bacteriuria into an infection. Abnormalities may be congenital or acquired, but in the context of IaI the most important is the passage of any device through the urethra (to cause a bacteriuria) plus even minor trauma (to offer the foothold). For patients undergoing urethral instrumentation a major risk factor is a pre-existing bacteriuria or UTI. In these circumstances up to 10% of patients suffer a bacteraemia. For most this is transient and symptomless, but in some cases the bacteraemia becomes a septicaemia. In this way a diagnostic cysto-scopy or a catheter change can be a lethal operation. Other than in an emergency, a patient who has or might have UTI should only be cystoscoped after bacteriological examination and, if necessary, treatment. In the absence of a laboratory or in great urgency, the investigation is done under the cover of prophylactic antimicrobials.

The incidence of iatrogenic UTI varies widely. Female sex, greater age and prolonged catheterization predispose and a major risk factor is the coexistence of any chronic urinary problem. Patients in these categories tend to suffer from repeated or protracted infections that are increasingly found to be due to multiply resistant Gram-negative rods, with *Klebsiella, Serratia* or *Pseudomonas aeruginosa* as examples. Ultimately *Candida* may appear and two or more pathogens may be present simultaneously. The mix of bacteria found in cases of UTI in one hospital are outlined in Figure 3.4 (p.39) and are described in more detail in Table 7.4 (p.152). They vary according to the way anti-microbials are used in the hospital concerned.

Some bacteria possess mechanisms ('**adhesins**') that help them to stick to surfaces. Such bacteria have an advantage in the urinary tract because they are less easily washed away by the flow of urine and so can more readily act as pathogens. This advantage is lost in catheter-ized patients, in whom bacteria without adhesins are more commonly found as pathogens. In catheterized patients another mechanism operates. The surface of the catheter acts as a foreign body on which a **biofilm** develops. This is a layer of sticky material derived from host body fluids into which bacteria, if present, are incorporated. These may contribute their own glutinous polymers to the film to build up a layer that resembles slimy blotting-paper in the interstices of which bacteria live and multiply, protected from body defences and antimi-crobial drugs. The encrustations that form may eventually block the lumen of a catheter. Not only do biofilms contribute to the develop-ment of infections but they also make them more difficult to treat.

Infecting microbes gain entry to the catheterized urinary tract at the points indicated in Figure 7.1. They may be introduced mechanically when the tip of the catheter passes through the colonized part of the urethra, at the time of catheterization. Once the catheter is in place they may travel through the space between the outside of it and the urethral mucosa (so-called **extraluminal spread**) or enter through the lumen of the catheter itself (**intraluminal spread**).

The presence of a catheter favours the abnormal multiplication of bacteria in the urethra. It is inevitable that, sooner or later, these bacteria reach the bladder, assisted by small reciprocal movements of the catheter generated by patient activity. If surgical tape is used to anchor the catheter to the patient's leg this not only prevents pressure by the self-retaining balloon on the neck of the bladder, but it also helps to minimize this movement. The bacteria that gain access by this extraluminal route, together with those introduced at the time of catheterization, are likely to have originated in the patient's normal flora, perhaps modified by hospital stay (p.44). Most of these iatrogenic UTIs are self-infections, perhaps of the autogenous type (p.48). Bacteria gain entry by the intraluminal route by retrograde spread from a colonization of the urine in the catheter bag. This is likely to have originated from the hands of staff, so cross-infection is a major element in the development of this variety of UTI. When clusters of infections due to indistinguishable resistant strains of bacteria appear, such cross-infection is the most likely cause of the outbreak. The relative importance of the extra- and intraluminal routes as causes of UTI varies with the quality of catheter care and the sex of the patient. In well-managed situations in females about 80% is thought to be due to extraluminal spread. The proportion is lower in males because bacteria must travel much further by the extraluminal route before they reach the bladder.

Diagnosis and treatment

Once bacteria reach the bladder they multiply and colonize the urine. This causes an (initially) symptomless bacteriuria which in catheterized patients differs from the transient form seen in uncatheterized individuals because it persists for as long as the catheter remains in place. The numerical criteria conventionally applied to bacterial counts in the diagnosis of UTI depend on the retention of urine in the bladder, with normal intermittent voiding. With a catheter in place voiding is continuous, with a small residual pool of urine round the balloon in the bladder neck. In these circumstances the standard criteria do not apply. If informed that patients are catheterized, many laboratory staff use a lower figure to separate patients who are likely to be bacteriuric from those in whom the count indicates no more than a contamination.

As bacteria continue to multiply in urine in the bladder, a colonization eventually turns into an infection. The point of transition from one state to the other is obscured in catheterized patients who cannot experience frequency and dysuria, the usual early symptoms of UTI. Perhaps only 20–30% of patients with catheter-associated bacteriuria develop symptoms, though this varies with the duration of catheterization. Symptomatic patients need to be treated, of course, but what of the others? Transient bacteriuria is not treated in uncatheterized patients and a view current among many clinicians is that, as catheter-related bacteriuria nearly always resolves when the catheter is removed, the same should apply. The situations are far from parallel, of course, but it is difficult to devise a solution that does not greatly increase unnecessary antimicrobial usage, itself to be discouraged. In the absence of symptoms the decision to use antimicrobials depends on risk assessment. The presence of neutropenia, other immune deficiency or anything else that increases the likelihood of septicaemia are factors to be taken into account. Unfortunately the usual result of

---

Box 7.2 A lethal catheter

Day

1 An active, well-oriented lady aged 72 was admitted to hospital with a six-week history of lower abdominal discomfort and distension with incontinence, frequency and nocturia. Her bladder was palpable. She was catheterized and 700 ml of clear urine was drained. Culture of the urine was reported negative.

2, 3 Improved, comfortable.

4 Fell out of bed.

5 Seen by the social worker with a view to discharge.

6 Catheter removed, for discharge tomorrow.

7 Not passing urine, by 6.00 pm her bladder was once more palpable. A catheter was reinserted and 400 ml cloudy urine drained.

8 Now disoriented, with a low fever. In the evening commenced cefuroxime and gentamicin. Blood and urine were cultured (subsequently *Escherichia coli* resistant to ampicillin and trimethoprim was reported in both specimens).

9 In septic shock. She died at 6.30pm.

This history illustrates how difficult it is to diagnose urinary tract infections in catheterized patients and the occasional weakness of the 'do not culture, do not treat' approach to it. This lady was killed by her catheter.

---

treatment given while catheters are in place is to exchange the microbes initially present for others of more resistant kinds. A counter to the idea that, in general, treatment should be withheld is the observation that lethal septicaemic shock can complicate asymptomatic bacteriuria (Box 7.2). The decision to treat or not to treat hangs on a fine balance of probabilities. It is not easily made.

Control

Catheter-related UTI may be caused by self-infection or by cross-infection and the bacteria that gain entry do so by the extra- or intraluminal route. Strategies for preventing the entry of bacteria into the urinary tract differ according to which route is followed. The greater importance of the intraluminal route may be judged by noting the large difference between the rates of colonization and infection following open (100% after four days) and careful closed drainage (100% delayed to 30 days). Closed drainage reduces the chance of intraluminal spread, but it cannot influence the extraluminal route. If patients' urines are colonized too early after catheterization (with a difference in timing between the sexes), the most likely cause is intraluminal spread due to poor quality care and a failure to keep the drainage system closed. It may be necessary to break the connection between a catheter and its collection system to irrigate the bladder to remove blood clots. The need to do this can be reduced if a three-way catheter is used, but multi-lumen catheters facilitate the inadvertent conversion of a closed system into an open one. As a cardinal rule closed systems should not be opened for trivial reasons. Urine samples should be collected from the closed system with a syringe and needle and not by breaking the connection between catheter and bag and, of course, never from the drainage tap attached to the bag itself.

The only regular reason for opening the system is to drain the urine that has collected in the catheter bag. The intraluminal spread of bacteria into the bladder is often preceded by a colonization of this urine and this results from carelessness when emptying the bag. Emptying should be a careful, clean procedure. Hands should be washed before and after and a plastic apron worn if splashing is likely. The jug or vessel should be thoroughly washed and dried between each use or (better) put through a washer-disinfector if one of these is used for cleaning bedpans and urinals. Gloves are necessary if the bag is constructed so that hands are likely to be contaminated with urine. Draincocks should be designed to avoid this. A nurse who moves from bag to bag without disinfection of the jug and appropriate handcare is negligent. In hospitals carelessness in the handling of catheters and drainage bags is generally more common at night.

The addition of a disinfectant to collecting bags might prevent bacterial colonization of the urine in it and some systems have incorporated non-return valves to prevent the reflux of possibly contaminated urine from the bag into the bladder. Neither strategy has been effective in all trials and in any case these measures are only necessary if simple primary rules are broken. The bag should not be raised above the level of the urethra nor allowed to trail on the floor. If there is a risk of reflux (when a patient is being moved, for example) the tubing should be clamped temporarily (and, of course, unclamped afterwards). Catheterization is usually ended after a few days, so in most cases the bag does not need to be changed. In long-term catheterization catheters should be changed as determined by patient assessment and the collection system should be renewed at the same time.

In extraluminal spread the initial entry of bacteria into the bladder is limited if meticulous attention is paid to aseptic non-traumatic procedure at catheter insertion. The lubrication of the urethra with an anaesthetic jelly containing an antiseptic has been shown to reduce the rate of infection. Catheters of smaller diameters are less traumatic and offer a reduced surface area for the development of biofilm. Smaller self-retaining balloons minimize the volume of the permanent pool of residual urine retained at the bladder neck. Catheters made of silicone rubber appear to delay the development of abnormal bacterial colonization in the space between the catheter and the wall of the urethra and catheters coated with or made of materials possessing special physical or disinfectant properties may also limit the development of biofilm. In practice, however, the results of trials of these innovations have been inconclusive or they await final adjudication.

The regular use of disinfectants round the urinary meatus for as long as the catheter is in place appeals to logic, but the results of trials conflict. One possible reason for this is that the abnormal colonization of the urethra is most often initiated at catheter insertion. Disinfectants applied subsequently to the meatus do not reach the whole length of the urethra which, of course, cannot be sterilized prior to catheterization. Another important reason for a lack of agreement between the results of trials is described in Chapter 4, p.79. In practice, while catheters are in place, it is thought sufficient to wash the perineal area to a good standard of personal hygiene twice daily and after bowel movements. With non-ambulant patients perineal care with soap and water at a similar frequency or as determined by nursing assessment is substituted.

The incidence of hospital-acquired UTI can be reduced by avoiding catheterization whenever possible and by removing any catheters that are inserted at the earliest possible moment (Box 7.3). The use of catheters would diminish if they were only available on prescription. Those who catheterize for convenience need to be reminded that the insertion of a catheter carries a mortality (Box 7.1). A catheter should not be used only to collect a specimen of urine. Incontinence pads can replace some catheterization with benefit to infection rates, but their use is associated with disposal problems. The use of condom-like collection systems for males can reduce infections, but only if damage to and infection of the skin of the penis is avoided. Suprapubic drainage reduces the chances of infection by the urethral route, but may be too invasive for short-term use. Intermittent catheterization has been used to replace both short- and long-term catheterization. In good hands the technique has prevented infections, but care is necessary to avoid trauma to the urethra.

It should be possible to monitor colonizations or infections in catheterized patients and react appropriately if they are too high (Figure 7.2). The process of audit will involve some urine cultures, though

---

Box 7.3 Catheter care

Nothing can prevent catheter-associated urinary tract infection, but the application of simple rules will reduce their number. Good practice in this respect is summarized as follows:

- Use catheters only when strictly necessary.
- Remove them as soon as possible.
- Use strict aseptic, non-traumatic technique at all stages.
- Secure catheters to limit their movement.
- Maintain a closed system of drainage.

routine bacteriological testing is not cost-effective. A sampling system should be devised, concentrating on departments of the hospital with higher rates of UTI (Figure 3.3). As with any survey, the active involvement of the clinical staff helps to ensure that it is properly performed and that the findings are acted upon.

Long-term catheterization is always associated with bacteriuria and UTI and this is a major cause of mortality in patients with spinal injuries due to trauma or disease. The management of these cases is a specialized subject with a large literature to which reference should be made by those who work in this difficult area.

WOUND INFECTIONS

Surgical sepsis, what it is, how to assess it, how it is caused and how to reduce its incidence.

Introduction

Medical dictionaries define a wound as an injury to the body caused by physical means with disruption of the normal continuity of body structures. Skin and mucous membranes need not be involved. This rather narrow definition appears to be designed to satisfy forensic requirements. The broader definition used here encompasses the results of any accidental or deliberate trauma that breaks the surface of skin or a mucous membrane. Surgical wounds, pressure sores and ulcers due to vascular or neural pathology and the results of thermal injury are included. The points of entry of tubes or wires into otherwise closed body cavities or the cardiovascular system are also wounds but, as these are considered in Chapter 7, p.156, and Chapter 8, p.182, they are not dealt with here.

Accidental wounds, burns and pressure sores may be treated surgically, so join operative wounds as 'surgical'. To aggregate them in this way is not as harmless as it seems. It has led to the inappropriate and expensive extension of practices thought to be relevant in the treatment of one kind of wound to the treatment of others (Colebrook, Duncan and Ross, 1948; Box 7.4). In the context of iatrogenic infections it is important to distinguish between wounds in which subcutaneous tissues are exposed to the outside world for days or weeks and those where the exposure is limited to minutes or hours. In the first category are burns, pressure sores, traumatic or other wounds with loss of superficial tissue and surgical wounds that result from the treatment of missile or blast injuries in which skin closure is deliberately postponed (delayed primary suture). In the second category are the wounds that arise from the great mass of elective and emergency surgery. The differences between them are fundamental.

Box 7.4   A doctor's dilemma?

In his 1948 paper Leonard Colebrook (1883–1977) wrote: '. . . a little reflection should have shown the surgeons that the infections of burns had similar origins to those of the operating theatre and might . . . yield . . . to a similar mode of attack'. He made this statement in the context of burns dressed in a room supplied with filtered air. He claimed that this, with other measures, eliminated most sepsis in burns. His conclusion was epidemiologically flawed and hopelessly over-optimistic. The introduction of penicillin may have had something to do with the apparent success. Colebrook was influential (he had been concerned with the introduction of flame-resistant nightwear for children) and his ideas were accepted. There is little doubt that his conclusions have influenced the design of operating rooms.

  As his pupil (and biographer), Colebrook must have been involved with Almoth Wright's attempt to 'stimulate the phagocyte', so he should have known better! (Almoth Wright is caricatured in George Bernard Shaw's play, 'The Doctor's Dilemma'.)

Pathogenesis

Infection implies the entry of microbes or their toxic products into tissues from which they are normally excluded, accompanied by some reaction from the host. The first lines of defence against such invasions are the skin and mucous membranes. When these are breached an invader encounters a series of secondary defences that depend on phagocytic and other cells that act in concert with a variety of humoral or chemical factors. The success or failure of these immunological mechanisms determines whether or not the exposed tissue is colonized and infected. Immune mechanisms may be defective. A general defect involves the whole body and it may have been imposed by medical treatment. Local defects follow when a discrete part of an individual's body is starved of otherwise competent immune factors. This may be the result of vascular or neural injury or disease, the presence of a haematoma, dead tissue or a foreign body. Local immune defects may also complicate medical or surgical treatments. In either case the compromised tissues respond with reduced efficiency to a microbial invasion. In these circumstances an intrusion that would be brushed aside by a normal person is more likely to result in an infection. This is the basis of the fundamental difference between the two types of wound defined above.

  In the early 1940s haemolytic streptococcal infections of burns, wartime wounds treated by delayed primary suture and industrial injuries, were subjects of intense study. In many cases large areas of more or less devitalized tissue were exposed for long periods. Such wounds were perfect culture media for streptococci, at that time the most feared of hospital pathogens. Colonizations, infections and in some cases septicaemia followed in progression. Once introduced into units where such patients were nursed, cross-infection ensured that every wound quickly became home to an enormous population of streptococci. From these sites the bacteria were available to contaminate the hands of staff and the environment, to create a vicious circle. It was a matter of observation that such wounds became infected after initial surgery, in the ward (Miles, 1944). In the case of burns investigators felt that infection was being transmitted through the air as well as by physical contact with members of staff. Filtration of the air was introduced to control the airborne route and the 'no-touch' dressing technique was employed to control contact spread (Colebrook, 1955). Quantities of disinfectants were used to decontaminate the environment .

In this situation why did simple handwashing, perhaps with a disinfectant, fail to duplicate the success enjoyed by Semmelweis almost a century earlier (Chapter 2)? The answer is that, by touch or by falling out of the air, streptococci reached devitalized surfaces that lacked their normal defences, including polymorphonuclear leucocytes (PML; p.51). Such a surface resembles a permanently open petri-dish of blood agar maintained at the perfect temperature of a bacteriological incubator. Cocci that may have already lost their virulence due to drying, often in numbers too small to overcome the PML of a normal person, were able to gain a foothold and multiply to produce a colonization. This soon became an infection as the initial invaders grew into a large population of now fully virulent streptococci. The uterine wounds of Semmelweis' patients were inaccessible to airborne microbes and did not lack PML (Chapter 2). This accounts for his success when disinfection of the hands removed large numbers of fresh, fully virulent streptococci.

The sequence of events described for streptococci seems also to apply to staphylococci and other bacterial pathogens. This explains why even the most elaborate care fails to prevent the colonization and infection of wounds in a burns unit (p.224). Burns, wounds treated by delayed primary suture, wounds closed after inadequate debridement and pressure sores are all special cases. When wounds of these types are concentrated in a ward or unit they are quickly colonized (and later infected) as a result of cross-infection or even by environmental infection. The bacteria involved in these units tend to be indistinguishable from each other and may be of little or no importance as pathogens in other parts of the hospital.

None of this is true of ordinary surgical wounds, where tissues are exposed to the outside world only briefly and in which the patient's innate system of immunity may be fully functional. Infections are much less common in these wounds and even in the same unit at the same time, the bacteria that cause them are often different. Infected surgical wounds of this type are distributed in hospitals as shown in Figure 3.3 and the bacteria characteristically involved are as outlined in Figure 3.4 and given in greater detail in Table 7.4. The single most common pathogen in surgical wound infections (SWIs) is *Staphylococcus aureus*. The data used in these figures and the table came from a hospital in which staphylococci were the cause of 40% of SWIs, divided between ordinary *Staph.*

aureus (22%) and the methicillin-resistant variety (18%). Gram-negative rods were judged to have been responsible in another 36% of cases.

Prevention of infection

As with any infection (p.20) a complex 'soil and seed' relationship exists between patients' tissues and colonizing bacteria. At the end of an operation all surgical wounds contain some bacteria. When the wound is closed the question is, can these contaminating colonists overcome the local defences and establish an infection? The outcome turns on the balance between host resistance on one hand and bacterial numbers and virulence on the other. The number of bacteria that can be dealt with by local defences is very large (p.21), but this ability may be compromised by a number of factors, summarized in Table 7.1.

The soil

To a large extent the local factors listed in Table 7.1 are under the direct control of the surgeon. Hippocrates recommended that wounds be treated gently and at the beginning of the modern era of surgery Halsted propounded the basic principles of wound care. Evidence to be produced later shows that surgeons sometimes need to be reminded that complete haemostasis, preservation of a sufficient blood supply, removal of devitalized tissue, obliteration of dead space and wound closure without tension are important if infections are to be avoided. Foreign bodies in wounds reduce by a large factor the number of bacteria that can be handled by tissue defences. The more traditional of the surgical foreign bodies – sutures, ligatures and tissues destroyed by heavy-handedness or diathermy – have been joined by a growing array of novel spare parts made of plastic, metal or animal tissues.

In an emergency there may be no time to correct any of the general factors that make the soil of wounds more fertile and so

Table 7.1 Factors on the patient's side of the equation that influence the development of a surgical wound infection ('the fertilization of the soil')

The soil
Subcutaneous tissue exposed through a wound at operation
-fertilized by:
- *local factors, in or near the wound,* many of them under the control of the surgeon: dead tissue, haematomas, poor circulation, foreign bodies (sutures, ligatures, prostheses, etc.)
- *general factors,* patient's state of health or debility; immune deficiency
- *modification by prophylactic antimicrobials*

encourage the growth of the bacterial seed (Table 7.1). In the case of elective surgery infections already present can be treated and other risk factors, if susceptible to correction, can be dealt with. In many cases antimicrobial prophylaxis has a profound effect (p.104).

The seed

The microbial components of the equation must be added to the factors that alter the susceptibility of a wound to infection. If logic is to be applied to the prevention of SWI, it is necessary to analyse the sources and vectors of the pathogens concerned (p.44). The possible sources are listed in Table 7.2. Some people carry the most common pathogen, *Staph. aureus*, as a part of their normal flora. When *Staph. aureus* causes an infection it may have come from the patient, from their attendants or from some part of the environment contaminated by a carrier. The most probable route can be identified by an examination of the incidence of SWI following different operative procedures.

Classes of surgery

It has been customary to divide operations into classes in accordance with differences in the risk of infection associated with each (Box 7.5). Four classes are commonly employed, though up to five have been used at various times (Table 7.3; see the Appendix for definitions). In a series of some 60 000 surgical operations studied by the Canadian surgeon Peter Cruse over a period of ten years, clean operations were associated with a rate of SWI of 1.5%. The figure for clean-contaminated operations was 7.7% and contaminated and dirty operations were infected in 15.2% and 40.0% of cases, respectively (Cruse and Foord, 1980). In other studies in which operations were categorized in this way similarly graded rates of infection have been noted. The point about the Cruse study was that patients were allocated randomly among a group of operating rooms (ORs) in a hospital and postoperatively they were

---

Table 7.2 Possible sources of the microbes that cause surgical wound infections ('the supply of the seed')

The seed
  Microbial contamination of a wound arising from:
- the *normal or abnormal flora* of patients' skin and colonized hollow organs or abnormal colonizations or infections elsewhere, invading the wound directly or through the blood, at the time of operation or later
- the *bodies of members of staff*, before, during or after surgery
- the *environment*, before, during or after surgery

Box 7.5 Early audits and the reduction of surgical sepsis

In 1896 George Emerson Brewer, a surgeon, shocked his colleagues when he announced that over 30% of patients who underwent clean surgery in the New York City Hospital developed infections post-operatively. By 1915 he was able to record that the rate had fallen to 1.4%.

In 1935 another surgeon, Frank Meleney, used an anecdote to open his report on a nine-year survey of sepsis, again in clean wounds and also in New York. A senior surgeon, asked about postoperative sepsis in clean surgery, declared that if all minor cases were included, the figure might reach 2%. In the first year the real figure proved to be 15.8%. In the following years it fell from a high point of 17.7% to 5%.

Both surgeons knew that different types or classes of surgery were associated with very different rates of postoperative sepsis. This was why they surveyed 'clean' operations, because they felt that high or low rates of sepsis associated with this variety separated 'bad' from 'good' surgery. In these early studies the terms 'clean', 'septic' and 'infected' were not defined (Brewer, 1915; Meleney, 1935).

Table 7.3 The incidence of postoperative sepsis in three surveys, stratified by the class of surgery that caused the wound, with (*in one case) the effect of reclassification by the simplified risk index (SRI, see text), which incorporates the factor of patient variability (an SRI of 4 is impossible with clean surgery)

Class of surgery	Postoperative sepsis (%)		
	1980[a]	1985[b]	1991[c]
Clean	1.5	2.9*	2.1
Clean-contaminated	7.7	3.9	3.3
Contaminated	15.2	8.5	6.4
Dirty	40.0	12.6	7.1
*Simplified risk index	0	1	2	3
Clean (from 1985[b], above)	1.1	3.9	8.4	15.8

(a) Cruse and Foord, 1980; (b) Haley et al., 1985; (c) Culver et al., 1991.

distributed (again at random) to a set of surgical wards. In this way patients were treated similarly with respect to the exposure of their wounds to environmental infection and (with the exception of the surgeons) to the risks of cross-infection as well. How may the large differences in the incidence of infection between clean and the other classes of surgery be accounted for?

When examined in detail the results showed that the rates of infection were the same for each operating room and ward. As just noted, rates varied according to the class of operation, but it was found that they also differed within classes, between individual surgeons. This difference was much more dramatic when clean surgery only was considered and the different types of surgery (general, orthopaedic, etc.) were separated. It was found that some surgeons were regularly returning rates of SWI following clean operations three or more times that of the 'best' of their colleagues. This was true for all the types of surgery. The inevitable conclusion was that environmental infection and cross-infection from OR and ward staff (excluding the surgeon who operated) could only be responsible for a fraction of the lowest rate of infection scored by the 'best' surgeons. A further conclusion was that postoperative care made a negligible contribution to the development of SWI. Because the skin edges of a wound closed at the end of an ordinary operation seal together within a few hours sufficiently to prevent the entry of bacteria, this is not surprising. Drainage by closed vacuum methods does not alter this, though, of course, open drainage does. The observation that wounds were not infected in the ward allowed the hospital to simplify postoperative wound care and so save money (p.184).

The statement that most ordinary surgical wounds are impervious to the entry of bacteria after a few hours requires some explanation. Biological functions rarely have definable points at which something is untrue at one moment and true the next. Even death defies definition in such simple terms and so it is with surgical wounds. The reason is explained in physiology textbooks where the process of blood clotting, the starting point for wound healing, is described. In this case the clot forms in what, if the surgeon has followed Halsted's precepts, is an insensible gap between the opposing surfaces of the wound. Events follow one another in a cascade that converts the soluble fibrinogen of the blood into fibrin, an insoluble polymer initially laid down as a loose, weak mesh. This happens within minutes and in the following few hours the loose network consolidates and contracts to form a much firmer structure into which fibroblasts begin to grow and lay down collagen as a prelude to final healing. Provided the wound is not forcibly torn apart this progression renders a wound increasingly impervious to the entry of bacteria from the outside at a rate that proceeds most rapidly at the outset. There is no point at which a wound suddenly becomes impervious to anything, but the process is substantially complete once the clot has retracted and has been anchored in place by the first fibroblasts. The speed with which this happens varies according to circumstances, but in nearly every case 'within a few hours' is a fair estimate.

Once Cruse had largely eliminated general cross-infection and environmental infection as sources of the microbes that cause SWIs, there remained for consideration only the patient and their surgeon. If the patient is the source then the best point at which to reduce the risk may be assessed from Table 7.1. The item most readily modified under the local factors is surgical technique. When he discovered that there were large and significant differences in infection rates between surgeons, Cruse informed them individually of their sepsis rates, with the performance of the others (given anonymously) for comparison. There was an immediate and dramatic improvement and the overall sepsis rate in clean surgery was halved. In this way he showed that some surgeons were associated personally with unnecessarily high rates of sepsis. It is most unlikely that this was because, uniquely among OR staff, the surgeons were themselves the sources of the microbes that caused these excess infections. This left self-infections with patients' own microbes as the most common causes of SWI. It appears that some surgeons had inadvertently contributed to SWI by adding to the fertility of the soil rather than by supplying the seed. That this is applicable more generally has been confirmed by other studies (see, for example, Mishriki, Law and Jeffrey, 1990). 'Surgical. skill' is undoubtedly multifactorial, but in the context of IaI important components seem to be experience and the volume of surgery performed (Wenzel, 1995).

Risk factors

The division of surgical operations into classes by Cruse and others has defined the crucial role of the surgeon in the prevention of surgical sepsis. As a measure of the risk of sepsis in a particular case, however, the class of surgery entirely ignores the important contribution made by the patient. As a part of the Study on the Efficacy of Nosocomial Infection Control (the SENIC project; p.84) Haley and colleagues set out to correct this deficiency (Haley *et al.*, 1985). They combined the class of surgery with nine other risk factors and calculated an index that estimated the probability for each patient of developing a surgical wound infection. Multivariate analysis of the ten factors showed that they could be reduced to four with little loss of predictive power. The factors that remained were identified in answers to the following questions.

- Was the abdominal cavity opened?
- Did the operation last more than two hours?
- Did it fall into either the former contaminated or dirty class?
- Did the patient have three or more underlying diagnoses?

Answers were scored zero for each 'no' and one for each 'yes'. The sum of the four figures forms a five-step (zero to four) 'simplified risk index' (SRI). This index was found to be a much more powerful predictor of surgical sepsis than the method that used classes of surgery on their own (Table 7.3).

The SRI was originally devised in 1970 in the initial stage of the SENIC project. In 1975–1976 it was successfully validated in the second stage of the project with data derived from 59 352 surgical patients admitted to 338 hospitals. Forty-six percent of these patients scored an SRI of zero. An index of zero was associated with a postoperative sepsis rate of 1%. Rates of infection in the other four categories (and the proportion of patients that fell into each) were 3.4% (32%), 8.9% (16%), 17.2% (5%) and 27% (1%), respectively. In this way the technique identified a subset of a little over 50% of patients who suffered 90% of the surgical sepsis experienced by all of them.

The power of this deceptively simple system became clear when the rates of sepsis among patients whose surgery had been classified as 'clean' were recategorized using the SRI method. For SRIs of zero and three the sepsis rates were 1% and 16%, respectively (Table 7.3). Similar gradations were found when the other classes were examined in the same way. The apparent simplicity of the system hides the fact that each of the four risk factors is composed of a complex of many interrelated lesser ones ('multicolinearity').

A different version of the SRI has been proposed by Culver *et al.* (1991). As a measure of a patient's contribution to their 'patient risk index' (PRI), a figure derived from the American Society of Anesthesiologists' system for preoperative evaluation was substituted for the rather simpler 'three or more diagnoses' of the SRI. The 'more than two hours duration' of an operation was also elaborated to incorporate what was felt to be a better indicator of surgical competence. In each case the actual time taken was compared with the average time for 75% of similar operations and awarded a score of one or zero for times more or less than this. Abdominal operations were not scored separately. Other methods have been proposed for the calculation of the risk of infection following surgery.

Perioperative factors
Other important conclusions emerged from the Cruse study. Several of these agree with independent observations made by others. The longer a patient stays in hospital preoperatively, the greater the chance of postoperative wound infection. The use of adhesive plastic drapes has been associated with a higher rate of sepsis than accompanies the use of waterproof or water-repellent disposable or reusable ones. Skin disinfection is best done with an active antiseptic with a residual action: in the Cruse study chlorhexidine and povidone iodine were used in conjunction with alcohol. Alcohol is an excellent fast-acting disinfectant. It does not need to be applied in such quantities as to leave pools of flammable liquid that might be ignited by electrodiathermy and it should only be used on exposed skin, where it can evaporate (pp.122–123). Clothing, dressings or tourniquets wet with alcohol left in contact with skin can cause chemical burns. It should not be used near the eyes.

Cruse found that shaving the operative site with a razor was contraindicated. When electric clippers were used the rate of SWI was lower and a depilatory cream (or no shaving at all) was associated with the lowest rates of infection (see also Seropian and Reynolds, 1971; Alexander *et al.*, 1983; Mishriki, Law and Jeffrey,

1990). In fact there is no microbiological reason for preoperative shaving (p.46). Removal of hair in the immediate locality may be desirable to keep it out of the operative field and, if they are used, to allow adhesive dressings to stick to the skin.

Over 10% of pairs of gloves were found to be punctured postoperatively. Perhaps because surgeons 'scrubbed' with an antiseptic detergent with a residual action (chlorhexidine) none of the patients who were exposed to risk as a result developed a wound infection. Cruse showed that 'scrubbing' of the hands of surgeons and their assistants with a brush could be limited to the first case on a list and that the routine 'scrub' may be simplified without danger to patients. He also noted that a survey that does not include a 28-day follow-up of surgical wounds loses data because a significant number of infections declare themselves after the patient has been discharged from hospital. As hospital stay gets progressively shorter, follow-up of patients into the community is essential for the accurate audit of surgical sepsis (see later). Lynch *et al.* (1992) reported on a study of the effect of preoperative whole-body bathing with an antiseptic preparation and they reviewed a number of prior, conflicting reports. They concluded that the practice is not cost-effective. The importance of prophylactic antimicrobials has been discussed in Chapter 6, p.104.

Pathogenesis refined by these studies

Quantitative ideas help us to understand these findings. A study found that over half the wounds that contain more than 100 000 bacteria per gram of tissue at the time of closure became infected afterwards. Wounds that contained less than this number of bacteria healed without infection (Krizek and Robson, 1975). Body defences can overcome a large bacterial challenge. The number of bacteria on or in healthy skin varies from one part of the body to another, with an upper figure of 400 000 per cm^2 (p.46). About 20% of skin bacteria survive effective surgical disinfection, to leave a maximum of 80 000 per cm^2 at the beginning of surgery. The bacteria concerned will usually be members of the normal skin flora, perhaps modified by stay in hospital (p.22). *Staph. epidermidis* is therefore one of the less pathogenic though more universal contaminants. A proportion of people are carriers of the much more pathogenic *Staph. aureus* (p.46). When found on skin they usually inhabit hair follicles, just below the skin surface. Here they are secure from the effects of shaving and are protected from disinfectants. It can be imagined that the number of *Staph. aureus* that can penetrate into a wound from the skin of such a carrier might, before the end of surgery, multiply to exceed the watershed figure of 100 000 per gram of tissue.

This is much more likely to happen when patients are shaved preoperatively. Shaving inevitably causes minor abrasions in which any

Staph. aureus present will start to multiply, their number increasing as the gap widens between the time of shaving and the beginning of the operation. Of course, if a surgical wound involves a contaminated hollow viscus such as the gut, the number of microbes available to initiate an infection is likely to exceed many-fold the number needed and the bacterial species concerned are more likely to be Gram-negative rods or anaerobes. In elective operations involving the colon preoperative preparation of the bowel is designed to reduce the level of this contamination. Surgical invasion of such sites explains why operations defined as clean-contaminated and contaminated or operations in which the abdomen is opened have higher rates of infection than clean ones. These numerical considerations support the idea that most SWIs are self-infections.

Alternative sources of the microbes that cause SWI are the OR staff and the environment. It has already been noted that these can only cause something less than the lowest rate of SWI achieved in clean surgery. Staff carry bacteria in the same way as patients but other than in cases where prostheses are involved and given elementary care, it is difficult to imagine a route by which the number of bacteria necessary to cause an infection might be transferred to a wound. Even a glove puncture is unlikely to leak the large number required, particularly if an antiseptic surgical scrub with a residual action has been used beforehand.

Environmental and airborne routes of infection

There remains the environment. Bacteria that settle on inanimate surfaces usually adhere to them firmly, so unless the surface itself is put into a wound, they are harmless. This clearly separates the risks associated with unsterile surgical instruments from all other surfaces in an OR. Bacteria suspended in the air are different, as they are in a position to fall into a wound or onto exposed instruments or other items that are to be put into it. The only important source of bacteria found in the air of an OR is the bodies of its occupants (p.216). In a badly ventilated OR these might reach a count of 20 bacteria-carrying particles (mainly skin squames) per cubic foot ($706/m^3$). Particles fall out of the air at a rate that varies with their size, but in a bad OR about 100 average sized bacteria-carrying particles would settle every hour on an open wound measuring 10 in^2 ($65\,cm^2$). Many particles carry more than one bacterium, but most are of less pathogenic species, predominantly *Staph. epidermidis*. It is difficult to imagine airborne microbes reaching the critical level of 100 000 per gram during ordinary, non-prosthetic operating, even when those falling onto instruments are added. It is evident that the numerical difference between the potential for causing SWI by self-infection and by cross- or environmental infection is very large.

It is necessary to remember that cross- and environmental infections are closely connected in ORs as the microbes concerned both come from the same place. Effectively all airborne bacteria in an OR and by extension those on inanimate surfaces are derived from the bodies, principally the skin, of its occupants. The suits of woven fabric usually worn in ORs are not a barrier to the passage of microbes from the skin. This is due to their style, with large openings at various points, and to the material from which they are made. Fabric sufficiently closely woven and close-fitting round the openings as to block the passage of skin squames and the bacteria they carry tends to be uncomfortable to wear as it also limits the passage of perspiration.

What has been said so far applies to the great mass of general surgery in which infections, though regrettable and costly, do not often seriously prejudice a successful outcome. The outlook is good in clean surgery where, with care, infection rates as low as 1–2% are achievable. The higher rates of sepsis in other classes of surgery are improving with the help of carefully designed antimicrobial prophylaxis. However, as modern surgery becomes more complex and particularly where foreign materials are implanted in critical sites, even a 1–2% rate of infection is too much. An infection that loosens a hip replacement is a tragedy; an infection round a prosthetic heart valve may be lethal. In cases like this there is a strong stimulus to assess the real contribution to surgical sepsis of the cross- and environmental routes of infection to see if already low rates can be made even lower.

Although it dates back to 1959, the US National Research Council study of the effect of irradiating the air in ORs with ultraviolet (uV) light provides relevant information (National Research Council, 1964). In this study nearly 15 000 patients were operated on in ORs fitted with uV lamps designed to irradiate the air to kill airborne bacteria. Each OR had two lamps, one real, the other a dummy that emitted a blue light so staff did not know one from the other. The lamps were switched on alternately in a predetermined way. Operative procedures were classified according to the risk of infection, using five categories rather than the four employed by Cruse. The effect of the uV light was to reduce the count of airborne bacteria by 50–60%. Overall infection rates were 7.4% for operations done under uV and 7.5% when the dummy lamps were on. Only in the refined clean class (elective clean operations closed without drainage) was a significant difference noted: 2.9% infections under uV compared with 3.8% without it.

A more recent study was conducted in the UK by the Medical Research Council (Lidwell *et al.*, 1982). This involved just over 8000 operations for the implantation of hip or knee prostheses. These were done in conventional ORs or in ORs with ultra-clean air (laminar flow) ventilation. During some of the operations in the latter, additional all-enveloping impermeable exhaust-ventilated suits were worn by the operating team. These special measures reduced the number of airborne microbes by between 70% and 99%. Unfortunately the use of antimicrobial prophylaxis was not controlled in the trial. Deep surgical wound infections were recorded in 3.4% of patients who did not receive antimicrobial prophylaxis and who were operated on in conventional ORs. When operations were done under ultra-clean conditions this rate was halved and the rate was halved again (to 0.9%) when exhaust suits were used. Prophylactic antimicrobials on their own reduced the deep sepsis rate to 0.8%. In a retrospective attempt to untangle the effects of antimicrobials and clean air, the investigators excluded some of the contributors they had originally included in the study. With this retrospectively imposed selective approach the interpretation most favourable to ultra-clean air was that when this was combined with prophylaxis (without exhaust suits), there were 43 fewer cases of deep sepsis for each 10 000 operations done compared with operations performed in conventional ORs under prophylaxis. The interpretation using all the data without retrospective exclusions is that there would have been 11 fewer cases (Meers, 1983a, b).

Different positions are taken about the cost-effectiveness of providing the extra facilities necessary to produce improvements of this order and what those facilities should be. The conclusion is that in operations where relatively large areas of tissue are exposed for longer periods and foreign bodies are implanted, some airborne infection does take place, but that much of it may be prevented by antimicrobial prophylaxis. It is clear that the effect of an implant greatly reduces the number of bacteria needed to initiate an infection (p.21).

Whatever decision is reached, it would be a very expensive mistake to allow a conclusion favourable to ultra-clean air to influence general surgical practice. Because most operations are completed more quickly, with less exposed tissue and the addition of little extraneous foreign material, the very small contribution to IaI made by the airborne route in prosthetic surgery is not relevant.

It is a pity that the MRC trial did not include a group operated on in conventional ORs by surgical teams wearing exhaust suits. As the source of airborne bacteria in ORs is the human body, it is probable that wearing these suits would remove the need for ultra-clean air. In this context it is notable that the skin of the head and face is more heavily bacterially colonized than most other areas of the skin. It has been shown that the rubbing of masks increases the shedding of bacteria from the face (Schweizer, 1976), so it is likely that caps and the necklines of gowns do the same thing. Any larger particles that separate from or are rubbed off the faces and necks of surgeons and assistants are more likely to fall directly into the sterile field, particularly in the downward flow of air provided by many ultra-clean laminar flow systems (Hubble, 1996). In this respect the horizontal laminar flow system seems preferable (p.218).

Minimal access surgery

The last ten years have seen a major change in surgery as laparoscopic techniques have been developed. Initial enthusiasm has been tempered by more sober evaluation ('Is the laparoscopic bubble bursting?', Treacy and Johnson, 1995). The growing literature relates mainly to technique and general outcomes, with infection mentioned only in passing. The consensus, however, is that rates of infection appear to be about the same as for conventional surgery of the same type (White, 1991). If it emerges that infections after minimal access surgery truly are just as common as in traditional surgery the insignificance of the contribution of the environment, and particularly the aerial route, to the genesis of surgical wound infections is heavily underlined.

Conclusion

It is increasingly difficult to make fully controlled advances in the prevention of SWI. It can be deduced from Figure 4.2 (p.75) that control and test groups of prohibitive size are required for a trial designed to detect a 25% improvement in a sepsis rate that is already as low as 1.5%. The figure approaches 20 000 in each group or a total of 40 000 operations. A multicentre trial involving the number of surgeons needed to achieve this figure would almost certainly collapse under the weight of its own complexity. At various time some have yielded to the temptation to make deductions from experiments in which microbes are pursued into and out of the environment, with no regard paid to their relationship to real infection. Infection control is concerned primarily with the prevention of infections. There is no doubt that many expensive rituals, of which some practices in ORs are good examples (see Chapter 10, p.215), are the result of the mistaken application of data derived from experiments in which the presence of microbes was considered in isolation, with no count made of the infections they caused (Box 3.5).

The surveillance of surgical sepsis is complicated by the shortening of postoperative stay in hospital, a difficulty compounded by the growth of 'day surgery'. When audit does not extend into the community there has been a progressive 'loss' of cases of SWI that has now exceeded 60% (Mishriki, Law and Jeffrey, 1990; Lynch *et al.*, 1992; Byrne *et al.*, 1994). (See Chapter 4, p.71, for more details.)

RESPIRATORY TRACT INFECTIONS

Among iatrogenic infections the most difficult to diagnose and to control are often those of the lower respiratory tract. This section describes why this is so.

Anatomy and physiology

The respiratory tract is divided into two by the larynx. The upper part starts at the nostrils, where inspired air passes into the nasal cavity and the nasopharynx. It then travels through the oral and laryngeal parts of the pharynx, territory shared between the respiratory and alimentary tracts. This shared territory is important in the context of iatrogenic infections. Next the air passes through the vocal cords to enter the lower respiratory tract and on through the trachea, bronchi and bronchioles finally to reach the alveoli.

Apart from the area just inside the nostrils, the part shared with the alimentary canal and the alveoli themselves, the respiratory tract is lined with a ciliated columnar epithelium. The cells that form this epithelium line its inner surface with a mass of microscopic hair-like cilia. Scattered among the epithelial cells are mucus-secreting cells that lay a sticky carpet on top of the cilia. The cilia beat to and fro, strongly in one direction and limply in the other, in a co-ordinated fashion. The movement of the mucus is upwards towards the larynx from the lower part of the tract and downwards from the upper part. Mucus on this 'mucous escalator' arrives in the oral or laryngeal part of the pharynx, where it may be swallowed or expectorated (Figure 7.3).

It is often said that the lining of the respiratory tract functions to warm and humidify the inspired air. This is true, but it has another even more critical function. It serves to keep the lower part of the tract (and some of the upper) free of microbes, to produce some of the privileged surfaces that are so important in the context of IaI (see Chapter 3, p.19, and Figure 3.1).

The air contains enormous numbers of tiny particles. A shaft of sunlight in a darkened room is visible because of the light reflected by them. A variable, generally small proportion of these particles carry microbes, mostly bacteria, viruses and fungal spores. With the exception of fungal spores, it is unusual for microbes to travel on their own. One or more of them are carried on a raft of some other material or are enclosed inside a fragment of something else (p.45). This is why most of the particles that carry microbes, although still tiny, are very much larger than the microbes themselves. If the particles contained in the 15 000–20 000 litres of air we inhale every day were deposited in the respiratory tract, the lungs would soon silt up and infection and death would follow. The combination of tortuous air passages and the mucous carpet prevents this. As air travels round the bends the particles in it are flung against and stick to the mucus on the walls of the tract, in the manner of a car that goes off the road when a corner is taken too fast. Particles trapped in this way find themselves on the mucous escalator to be carried out of the respiratory tract at a speed of about 16 mm a minute.

Insect-sized particles are filtered out by hairs inside the nostrils.

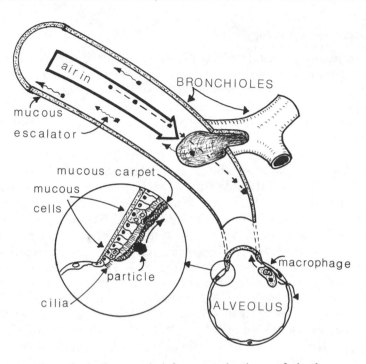

Figure 7.3 The principal natural defence mechanisms of the lower respiratory tract, with particles trapped by the mucous carpet and removed by the mucous escalator or dealt with by alveolar macrophages.

The mucus carpet of the upper tract accounts for nearly all those that measure more than 10 μ in diameter. Most smaller ones (and larger ones as well when breathing through the mouth) are caught in the trachea and upper bronchi. Smaller and lighter particles more readily 'take the bends': a light sports car can go round a corner more easily than a heavy lorry. More and more of the smaller particles are trapped as the air containing them travels down the bronchi and bronchioles. To reach the alveoli a particle must be smaller than 5 μ and ideally 1–2 μ in diameter. Particles that penetrate as far as this meet a new defender, the macrophage. These phagocytic cells engulf small particles and then remove them to the exterior by crawling out of the alveolus onto the mucous escalator or they carry them into the interior of the body by way of the lymphatic system. Particles smaller than 1 μ may make the round trip without striking any surface and are exhaled (Figure 7.3).

There are other defence mechanisms. Sneezing and coughing are reflexes triggered when larger particles enter the nose or land in the larynx, trachea and larger bronchi. In addition to the mucous escalator, the surface of the respiratory tract is protected by a variety of chemical and cellular elements that belong to both the innate and adaptive systems of immunity (p.20).

The pathogenesis of respiratory infections

As with its anatomy, infections of the respiratory tract are divided into upper and lower varieties. Infections may be confined to the upper respiratory tract (upper respiratory tract infections, URTI), begin in the upper part and spread into the lower (descending infections) or establish themselves from the outset as lower respiratory infections (LRIs). Infections of the upper tract may be trivial (a light cold) or very unpleasant (acute sinusitis) but do not often threaten the lives of healthy people. Many are caused by viruses. Some of these, including the respiratory syncytial and influenza viruses, may cause life-threatening descending infections in infancy and old age, respectively. These may be acquired in hospitals or nursing homes where they present as outbreaks of IaI.

Hospital-acquired LRI is a more serious threat because it can lead to potentially lethal pneumonia and septicaemia. Other than as just noted, infections of the upper tract are not a major feature of IaI, but bacterial pneumonia is an important component of it. To what extent LRIs are secondary to preliminary viral infections is not known. The frequency and distribution of LRI in hospitals is described in Chapter 3, pp.28 and 30, and in Figure 3.3. The latter illustrates how these infections are concentrated in intensive care units (ICUs), with a lower but still significant incidence in surgical and medical wards. In the latter they are associated with breathing difficulties after abdominal operations or the admission to medical wards of patients predisposed to infection because they are suffering from organic lung disease or from impaired respiration following a cerebrovascular accident.

Infections develop when the defences that protect the lower respiratory tract are compromised in some way. The cilia may be defective congenitally or their activity is suppressed by tobacco smoke, alcohol, the misuse of drugs and by some infections. Microbes that attack and destroy cilia are particularly harmful. Some viruses, mycoplasmas and *Bordetella pertussis*, the cause of whooping cough, do this rather well. Toxic agents may inhibit the cough reflex. As major contributors to IaI, the factors listed are joined by some therapeutic drugs among which the morphine group and anaesthetic agents are particularly active. Tracheal intubation (either for anaesthesia or to allow long-term mechanical respiratory support) bypasses the principal defence by interrupting the 'up' mucous escalator. The mucous carpet plus any particles that have adhered to it are trapped and the mixture accumulates in the lower part of the tract. This has to be sucked out mechanically by a process that is not only comparatively inefficient but also adds an opportunity for cross-infection. In addition, the heavily contaminated secretions that collect immediately above the balloon on cuffed endotracheal tubes may leak past it or run down into the lower tract when the balloon is deflated. Tracheal intubation is very

Table 7.4 The incidence of different forms of IaI per 1000 cases in a teaching hospital, 1986–1989 (Meers and Leong, 1990)

	Rate / 1000 cases of IaI
Urinary tract infections	412
Surgical wound infections	255
Lower respiratory infections	133
'Other' infections:	
IV cannula-associated infections	60
Septicaemias	42
Skin infections	28
Gastrointestinal infections	11
Infected pressure sores	5
All other infections	54

Key: IV, intravascular.

Table 7.5 The frequency of causative bacterial pathogens in septicaemia compared with the frequency in the more common kinds of HAI, from which the septicaemias probably originated (data from the NUH, Singapore, 1986–1989, Meers and Leong, 1990)

Pathogen	*Percent infections caused by the pathogens named*					
	Sept	*UTI*	*SWI*	*LRI*	*ICI*	*Skin*
Gram-negative rods						
Esch. coli	14	23	8	2	2	3
Klebsiella spp.	21	21	11	11	7	6
Ps. aeruginosa	3	7	8	19	4	4
Acinetobacter spp.	4	4	4	8	10	2
Proteus spp.	1	7	3	3	1	5
Enterobacter spp.	3	2	3	2	0	1
Staph. aureus						
ORSA	11	2	22	7	24	30
MRSA	11	3	18	25	26	29
Staph. epidermidis*	10	2	2	1	9	0
Streptococcus spp.	9	19	9	7	4	4
Candida spp.	6	4	3	8	8	10
Other pathogens	7	6	9	7	5	6

* *Staph. epidermidis* is named here in the usual clinical sense, to include all coagulase-negative staphylococci except *Staph. saprophyticus*.
Key: Sept, septicaemia; UTI, urinary tract infection; SWI, surgical wound infection; LRI, lower respiratory infection; ICI, intravascular cannula-associated infection; Skin, skin infection, ORSA, ordinarily resistant *Staph. aureus*; MRSA, methicillin-resistant *Staph. aureus*.

common in ICUs and this explains the heavy concentration of LRI in these units.

A summary of the microbial causes of LRI is given in Figure 3.4, with more detail in Table 7.5. In general Gram-negative rods

(GNRs) are the most frequent causes of hospital-acquired LRI. Between 1986 and 1989 in the hospital that was the source of the data presented in Figure 3.3 and Table 7.5, 45% of all LRIs were due to GNRs, among which *Pseudomonas aeruginosa* caused 19%, *Klebsiella* spp. 11% and others 15%. *Staphylococcus aureus* was next most frequent (32%), divided between 'ordinary' strains (7%) and methicillin-resistant *Staph. aureus* (MRSA) (25%). Other microbes were the cause in 23% of cases. Other centres have recorded higher incidences of infections due to GNRs, ranging up to 60%. In Singapore the greater frequency of intubation through a tracheostomy rather than by other routes may account for the higher incidence of infections due to *Staph. aureus* in the data presented (see below).

Gram-negative infections

The chain of events that leads to hospital-acquired LRI due to GNRs is well known. The normal flora of a healthy pharynx includes few enterobacterial GNRs, but they are more commonly found there in seriously ill patients. It seems that this is due, at least in part, to the loss of fibronectin, an important surface component of oropharyngeal cells. When present, this material encourages the adhesion of Gram-positive bacterial cells. In its absence, GNRs adhere more freely. The saliva of seriously ill patients is likely to contain an enzyme that destroys fibronectin. Such patients are also more likely to be treated with antimicrobial drugs. As a result their disordered oropharyngeal flora almost certainly includes multiresistant GNRs. A breakdown of respiratory defences allows these GNRs to enter, colonize and perhaps infect the lower respiratory tract. Tracheal intubation for mechanical ventilation is an excellent way of achieving this, so LRI due to GNRs is common in ICUs (Figure 3.3), where serious debility, extensive antimicrobial therapy and assisted respiration commonly coexist. It follows that LRI due to GNRs is often a self-infection of the autogenous kind (Figure 3.5). Cross-infection also plays a part because tracheal tubes are subjected to a lot of manipulation.

Gram-positive infections

In the hospital just mentioned MRSA was found more often than sensitive staphylococci as a cause of iatrogenic LRI. It is likely that this was due to cross-infection. A tracheostomy is a wound which, like any other wound, may be the site of a staphylococcal infection (p.137). This provides another source of bacteria which may gain access to the respiratory tract when the tracheostomy tube is handled, for example when the secretions trapped below the tube are sucked out or the tube is changed. Some surveys have shown that infections are more common in patients ventilated through a tracheostomy than when the oro- or nasopharyngeal routes are used.

Environmental infections

The other potential source of infecting microbes is the environment. Apart from the air itself, ventilators are particularly important. Humidification of inspired air is now usually with a humidifier rather than a nebulizer, so contaminated water is not actively sprayed into patients as was formerly sometimes the case. Any wet environment is a potential hazard, however and this one requires special attention. Humidification of the air as it is inspired and condensation from it as it is expired makes ventilator tubing wet, so able to support the growth of GNRs. The tubing needs to be changed at intervals. This should be done carefully, as a clean procedure, the hands being washed before and after. Opinions vary on the frequency of these changes and although the decision is somewhat arbitrary, it has financial consequences. A change every 24 hours was originally recommended, but it is now thought that this may safely be extended to 48 hours. Changes may be made even less frequently if heat-moisture exchangers are used and sometimes in the case of neonates (Cadwallader, Bradley and Ayliffe, 1990).

Other risk factors

Other risk factors associated with a higher incidence of LRI are age, immunosuppression and surgery. Each time the endotracheal balloon is deflated heavily contaminated secretions trapped above it slide into the lower part of the tract from where the interrupted mucous escalator cannot remove them. Another risk factor is the use of cimetidine or ranitidine to suppress gastric acidity to prevent the development of stress ulcers and gastrointestinal bleeding. Gastric acidity is the normal defence mechanism that keeps the stomach almost free of bacteria and protects the intestines from invasion by organisms present in food. In the absence of acid the stomach is colonized by a variety of bacteria that may regurgitate through the oesophagus into the pharynx and so reach the larynx. This is particularly likely in comatose or semicomatose patients with nasogastric tubes in place. The use of sucralfate has been recommended to protect the gastrointestinal mucosa from injury in a way that does not inhibit or neutralize gastric acid.

Diagnosis

The diagnosis of hospital-acquired LRI is often difficult. In intubated patients the progression over a period perhaps measured in days from an inevitable colonization of the trachea by potentially pathogenic bacteria to tracheobronchitis and then to pneumonia means that bacteriological criteria on their own are of little use. The usual clinical features of a respiratory infection, a cough and the production of sputum cannot apply in intubated patients, who may be febrile or have raised white cell counts for other reasons. X-rays may not help. Post-mortem studies have shown that the ante-mortem diagnosis of this form of IaI is very unreliable. The problem is that the best diagnostic methods (transtracheal aspiration, bronchoscopic examination, needle or open lung biopsy) cannot easily be applied to patients attached to mechanical ventilators or carry an unacceptably high morbidity in seriously ill patients or both.

Prevention

It will be supposed from this doleful litany that the control of iatrogenic LRI is far from easy. It is indeed a serious problem. It was noted above that, eventually, urinary tract infections are inevitable in catheterized patients. The microbes concerned gain access by intra- or extraluminal routes. The same applies in the case of respiratory intubation. The intraluminal route via the inhaled air cannot be controlled unless patients breathe sterile air, a thing not easily achieved and in any case of lesser importance. The proportionately much more important extraluminal route is even less controllable. With intravenous, and to a lesser extent urinary catheters, emphasis is placed on the microbiological environment at, and the care of the entry point of the catheter. In the case of respiratory intubation the tube passes through or very close to the oropharynx which is very heavily colonized with potential pathogens even in its normal state, and significantly more so in severely debilitated patients who need to be intubated. When the 'artificial' or 'iron lung' was used for the support of victims of poliomyelitis whose respiratory muscles were paralysed, the patients did not need to be intubated and LRI was less common. There is no doubt that, in the case of modern respiratory support, the endotracheal tube must be blamed as a major cause of IaI. Despite procedures and policies designed to limit cross-infection (for example, wearing gloves and scrupulous aseptic technique when caring for the tube and sucking out secretions – see below), infection is, eventually, inevitable. They are, in the main, autogenous self-infections secondary to colonizations that may have been initiated as a result of cross-infection. The best that can be done is to delay their onset and then to treat them as they arise.

The outlook is less bleak for the prevention of infections among patients who are intubated briefly for anaesthesia during surgery. In these cases simple preventive measures such as early postoperative mobilization, breathing excercises and physiotherapy are usually successful.

The care of suction equipment

The source of the suction required to remove secretions may be a permanent, built-in piped supply or a free-standing pump. To avoid contamination of mechanical components or permanent pipework, reusable or disposable collecting vessels and filters are interposed between the suction catheter and the source of the negative pressure. While the system is in use an antifoaming agent, but not a disinfectant, may be added to this vessel. A disinfectant may be sucked into it afterwards if the material it contains is thought to be seriously hazardous. Non-disposable collecting systems are washed out, disinfected (in a washer-disinfector if one is available) and stored dry until needed again. This process is carried out not less often than daily when suction apparatus is in constant use. The catheters should be disposable and used once or, if repeatedly, only on the same patient and for not more than 24 hours.

OTHER INFECTIONS

Iatrogenic infections associated with intravascular cannulation and iatrogenic septicaemia, skin infections, gastrointestinal infections, eye infections and infections associated with implants made of plastic are discussed.

Introduction

'Other' infections are those left when the three more common varieties of IaI (urinary, surgical wound and lower respiratory infections) have been dealt with. They vary in numbers and types from survey to survey, according to the definitions used, the types of patients concerned and whether or not mild upper respiratory infections, for example, are included. Table 3.2 shows that in 13 large surveys 'other' infections averaged 23% of all those that were hospital acquired. This figure is quite close to the 20% found between 1986 and 1989 in a teaching hospital in Singapore (Table 3.3) so despite variations between surveys, these data are presented here for illustrative purposes. Table 7.4 compares the incidence of the more numerous of the 'other' infections with the three more common varieties. They are discussed in the order of their frequency.

Infections associated with intravascular cannulation

Diagnostic or therapeutic invasions of patients' cardiovascular systems are frequent events. Here we are concerned with cannulation, a procedure that involves the placement of metal or plastic tubes into veins or arteries for periods of a few hours to weeks or months. Well over half the patients admitted to hospitals may be cannulated in this way. If the brief venepunctures made for the collection of blood samples and other diagnostic or therapeutic manoeuvres are included, the proportion approaches 100%.

The tubes involved are peripheral or central venous catheters including steel needles, cuffed catheters for long-term use, catheters for the delivery of total parenteral nutrition (TPN), balloon-tipped catheters used for sampling and monitoring in the central pulmonary circulation and arterial and umbilical catheters. All of these are subject to the same problems of infection to an extent that depends on their diameter, the length of the intravascular portion, the material from which they are made and the chemical nature of the material, if any, being infused and the length of time they are left in place. Cannulas inserted through areas of skin with a more abundant normal flora (particularly in the groin) or into the umbilicus are more likely to be associated with an infection than those inserted at other sites.

Vascular cannulation provides a high road along which microbes may pass to enter the cardiovascular system. Children and younger adults who are not severely immunocompromised can overcome a surprisingly large microbial challenge by this route, but this ability is diminished in infants, the elderly and the immunocompromised. As with urinary catheters and endotracheal tubes (pp.130 and 155), bacteria gain access to the vascular system by two routes: intraluminal and extraluminal (see also p.182).

Intraluminal infection follows when microbes gain access to fluids before or during infusion or transfusion or are present in the fluid in pressure-monitoring devices. This may happen for a variety of reasons. These range from failures during the manufacture of infusion fluids to punctures or cracks in their containers, accidents in making additions to infusions in hospital pharmacies, wards or departments or as a result of contamination gaining access through the various joints in or the supplementary injection ports of giving sets or, finally, following the attachment of piggy-back containers. Bacteria are able to multiply in most infusion fluids. The number that gain access initially is usually small and unimportant, but given time a trivial contamination turns into a lethal one. The longer the interval between the contamination of a fluid and its infusion, the greater the danger. There have been a number of well-publicized incidents in which manufacturing failures have led, months later, to the use of containers of contaminated fluid, with consequences that have involved numbers of patients (Anon, 1974; Box 6.1). Contamination nearer the time of infusion usually involves small groups or single containers of fluid and the hands of staff are likely to be the source of it. This type of contamination is the origin of sporadic cases of septicaemia the source of which may not be identified.

Extraluminal infections are self-infections or cross-infections by the autogenous route. Cannulas are inserted through the skin or via an umbilical stump. Neither site can be sterilized. The introduction of the cannula is commonly percutaneous or it may be by surgical cut-down. Unless gloves are worn, the equally unsterilizable hands of the operator making the insertion may be the source of contamination. Other sources are hands that manipulate the infusion apparatus later or, worse, fiddle with it unnecessarily. After a cannula has been inserted the outer surface of its intravascular portion develops a coating of fibrin plus other materials to form a biofilm sheath which, at the site of vascular puncture, is in contact with extravascular tissues and the surface of the body.

Bacteria derived from the patient's skin or the hands of members of staff, multiply at the site of epidermal puncture. The establishment of an abnormal colonization at this site is abetted by the presence of the inevitable foreign body and the nutritious fluid that oozes from the puncture wound. In time the colonization becomes an infection, but in either case the bacteria responsible are likely to spread round the outside of the cannula, to penetrate skin and subcutaneous tissues to reach its intravascular portion. Here they multiply in the biofilm, protected from the body's defences. To aid this process some bacteria possess adhesins that allow them to adhere to the material of which the cannula is made. As multiplication proceeds bacteria that reach the surface of the biofilm stimulate the deposition of more fibrin and they may be shed into the bloodstream in the form of infectious emboli. The more extensive the intravascular part of the cannula and the longer the duration of cannulation, the greater the hazard. If the tip of a colonized cannula lies close to or in the heart more serious complications may develop as endocardial or valvular damage predisposes to infections at these sites.

As with urinary catheters, the development of a biofilm may be delayed if intravascular cannulas are made of more sophisticated materials (pp.133 and 183). Over the years the diameters of cannulas have been reduced, so they present a smaller surface area for these undesirable changes. Steel needles are short and less likely to encourage adhesion of fibrin or bacteria, so these older alternatives to plastic cannulas are safer for short-term infusions in peripheral veins. Unfortunately if they are not inserted skilfully and properly cared for, their use may lead to extravasation of infusion fluids into perivascular tissues.

Inflammation of a vein is **phlebitis**. Inflammation may be due to physical or chemical trauma as well as to an infection. The physical trauma of skilful venepuncture is negligible, but inexpert puncture is not and in any case the long-term presence of a foreign body in a blood vessel inevitably causes some damage. Many of the fluids infused through cannulas also damage the lining of blood vessels. Some drugs given intravenously are so irritant that they must be diluted before infusion. Solutions for TPN are very corrosive and cannot be diluted sufficiently to allow them to be given into a peripheral vein. To overcome this they are administered directly into a central vein to be diluted directly into a large volume of rapidly flowing blood. The first inflammatory changes of cannula-related phlebitis may be due to non-infectious trauma. Bacteria that colonize the cannula track eventually convert this initially sterile inflammation into an infection. As with other forms of iatrogenic infection, the precise point at which a colonization becomes an infection cannot be determined. In its ultimate state an infectious phlebitis converts a vein into an intravascular abscess. This is rare, difficult to diagnose and potentially lethal. A patient with a persistent septicaemia who fails to respond to apparently adequate antimicrobial treatment may be suffering from suppurative phlebitis. This can be a special problem in patients with burns.

Authorities agree that most cannula-related sepsis is of the sporadic extraluminal type. The less common intraluminal infections due to the infusion of fluid from multiple containers contaminated at source appear as clusters of septicaemias involving the same, perhaps unusual pathogen. Device-related infections (like their counterparts in catheter-related urinary infections – p.126) tend to be associated with what have been termed 'line violations', in which closed systems are unnecessarily or carelessly broken into.

The bacteria typically responsible for cannula-related infections are shown in Table 7.5. Staphylococci cause well over half of them. These normal inhabitants of the skin cause self-infections, perhaps of the autogenous sort, or cross-infections. Gram-negative bacteria may colonize the skin when it has been allowed to become wet and unhealthy, so that infections with these moisture-loving bacteria also develop as a result of self- or cross-infections. Other potential sources are infections elsewhere in the patient. In this case the bacteria responsible may reach the cannula through the blood itself (haematogenous spread).

The prevention of cannula-related sepsis

The longer an intravenous line is in place, the greater the chance that the inevitable abnormal colonization at the puncture site will develop into an infection. Studies have shown that the number of infections begin to rise significantly after a cannula has been in place for 72 hours. This is why external lines should be changed at this point. If possible the cannula and the site of cannulation are changed at the same time. This is often difficult and in the case of central lines that must be maintained for long periods, impossible. Some central venous catheters for extended use are designed with cuffs that encourage the growth of tissue cells into them. The object is to form a living barrier to prevent the inevitable bacterial colonization at the puncture site from reaching the vein round the outside of the cannula (extraluminal spread).

Other methods that have been applied in an attempt to prevent cannula-related sepsis are the use of antiseptic ointments or occlusive dressings at the venepuncture site. Trials of antiseptics have given inconclusive results. As these agents cannot sterilize the skin, this is not surprising. Some occlusive dressings, particularly of an adhesive and perhaps transparent variety, have been shown to encourage an overgrowth of skin bacteria under them. They do this by providing the warm, moist environment bacteria love. However, they do keep fingers at bay. The results of controlled trials of their use are not consistent. One reason for this may be that while both antiseptics and occlusive dressings might reduce the sepsis that results from crossinfections, neither can do much to prevent self-infections. Differences in procedures and the quality of care determine that the proportion of cannula-related infections attributable to intra- and extraluminal spread varies between hospitals. A trial done in a hospital where crossinfections contribute significantly to the total of cannula-related sepsis may show that antiseptics and occlusive dressings are effective. The opposite result is likely where procedures that minimize the possibility of cross-infection are already in place (see Maki, 1994; Parras *et al.*, 1994).

Septicaemia

In this book the term 'septicaemia' is used to describe the presence of bacteria in the blood when this is accompanied by symptoms. Its symptomless counterpart is bacteraemia. Septicaemia is a common terminal event in severe infections. Except when bacteria are delivered through a cannula directly from an external source, it must be inferred that septicaemias originate in pre-existing infections somewhere else in the body of the patient concerned. In the context of IaI septicaemia is most often secondary to urinary, surgical wound or lower respiratory infections or to extraluminal cannula-related sepsis. The state of the patient's immune system determines how severe a primary infection must be to be able to cause septicaemia. In severe immunodeficiency septicaemias may arise from primary infections so inconspicuous that they cannot be located.

In Table 7.5 the bacteria that were identified in a teaching hospital as the causes of hospital-acquired septicaemia are named. The bacterial causes of some primary infections that might have been the sources of the septicaemias are listed for comparison. There is an obvious connection between the distribution of bacterial pathogens found in cases of septicaemia and the four most frequent kinds of IaI. As septicaemias are always secondary to infections or contaminations somewhere else, their control rests entirely on the control of these primary sources.

Skin infections

For a significant part of the population *Staphylococcus aureus* is a member of the normal flora of the skin (Table 3.4). It is also the principal cause of hospital-acquired skin infections (Table 7.5). Table 3.5 shows that proportionately, staphylococci cause more IaIs in infants than in older children or adults. The figures are 48% in the neonatal intensive (or special) care unit and 56% in the neonatal unit (together 51%) compared with 22% for all other areas of the hospital. Figure 3.3 indicates that 'other' infections are also found most commonly in the same units. Most of them were skin infections. The reason for this is that infants, born sterile, begin to develop their skin flora immediately they enter the world. Their microbiologically virgin surfaces must come to terms with *Staph. aureus*. Some infants achieve this without disease, but a significant number develop infections. Most are minor, but they need to be kept to a minimum, for two reasons. First, when the incidence of staphylococcal colonizations in neonatal units rises above about 75%, cases of more serious staphylococcal sepsis begin to appear. The relationship between rates of colonization with *Staph. aureus* and the incidence of infections caused by it is referred to again on p.165. The second reason is that mothers who develop breast abscesses are nearly always infected with resistant hospital staphylococci acquired from their infants.

Gastrointestinal infections

The small number of gastrointestinal infections recorded in Table 7.4 were due to rotaviruses, *Clostridium difficile*, some other cases of antibiotic-associated diarrhoea and a few in which heavy growths of *Aeromonas* spp. were interpreted as infections. There was one hospital-acquired infection due to a food-poisoning salmonella. This was in an infant born to a mother who was just recovering from an infection with the same strain. This distribution of pathogens is not uncommon in gastrointestinal infections acquired in acute hospitals, though in more recent years there has been a major increase in the incidence of infections with *C. difficile*. In some places salmonella IaIs are more common and they may be endemic. Otherwise they appear as causes of outbreaks of infection.

Salmonella food poisoning in the community is usually acquired from food contaminated at source. At some stage it has then been handled improperly so that a small innocuous dose of salmonellas is converted into a large infectious one, in the order of millions in a helping of contaminated foodstuff. This large dose is necessary if some are to survive the acid in the stomach and reach the intestine, to start an infection. Depending on the number of people who eat the contaminated food, the result will be a small or large outbreak of food poisoning. Secondary spread from person to person is uncommon but may be seen within families, particularly where there are children, and in hospitals.

Salmonellas may be introduced into a hospital in the same way as happens in the community or a patient may be admitted who is

already suffering from a salmonella infection. What happens next makes hospital outbreaks of salmonellosis different from those in the community. In long-stay hospitals and nursing homes where mentally disturbed or retarded children or adults are cared for, it is possible to transmit the large infecting dose directly by the faecal–oral route. This cross-infection is the result of difficult hygienic conditions in these places, particularly when patients are suffering from diarrhoea. The situation is worse in psychogeriatric units because less gastric acid is produced in old age. This allows a smaller dose of salmonellas to start an infection, so person-to-person transfer (cross-infection) happens more easily.

In acute hospitals the very large numbers of salmonellas needed to cause infections in healthy people are less likely to be transferred directly from one person to another. Patients in hospitals are far from normal, however and their secretion of gastric acid may be deficient for a number of reasons. Patients with low gastric acidity are susceptible to infection when they ingest much smaller numbers of salmonellas. Epidemiological evidence suggests that infants may be infected by the small numbers of salmonellas carried on lightly contaminated hands. The same thing seems to apply to debilitated adults or those taking antacid treatment. When a number of patients who fall into one of these categories are concentrated in one part of a hospital, any salmonella that is introduced easily becomes entrenched. A serious cross-infection problem then develops as newly admitted, uninfected patients provide the fuel that keeps the cycle of infection turning and sooner or later members of staff will be involved. It should be possible to control the situation by scrupulous attention to handwashing (staff and patients), but this may not be enforceable. It may be necessary to close the ward or wards concerned to new admissions until the salmonella has burnt itself out and the last carrier patient has been discharged or the last carrier member of staff has been cleared (Box 7.6).

Viral diarrhoea and vomiting may be severe and can disrupt activities in medical establishments. Outbreaks are more common among children. They are more often caused by rotaviruses, but the Norwalk group of viruses and some adenoviruses, astroviruses, caliciviruses, coronaviruses and small round structured viruses have also been implicated. These agents are transmitted by the hand-to-mouth, faecal–oral route or via food. In the context of modern paediatric units, outbreaks are difficult to control. Members of staff are often involved.

The residual infections

These include small numbers of cases of IaI, of many different kinds. With two exceptions, they are sufficiently like their community-acquired equivalents to require no further comment, except to note that they may be caused by hospital strains of bacteria.

Box 7.6 Salmonella sources and reservoirs

One September, a patient developed diarrhoea with fever and abdominal pain ten days after major surgery. *Salmonella senftenburg* was found in her faeces. She was isolated in a side ward. In the next four weeks three more patients in the same ward were infected with the same salmonella. The ward was closed to new admissions for seven days and the stools of all patients and many of the staff, including some of the doctors, were examined. Two patients (who remained on the ward) were found to be infected.

Early in December when the ward reopened a new patient developed diarrhoea three days after admission. In the next week seven more patients and 18 members of staff were found to be carrying the salmonella. Most were symptomless or only mildly upset, though some junior nurses were quite ill. The ward was closed again and infected patients who were unfit for discharge were sent to an isolation unit. The environment of the ward was examined and the salmonella was found in the sluice, in a bath, on three sheepskins and on commodes. Extensive environmental cleaning was instituted. The ward was reopened in January.

In all, 14 patients and 18 members of staff were infected, either clinically or subclinically. One nurse continued to carry the organism and did not work for three months, the remainder were off work for between two and four weeks. The three sheepskins were burnt and all the linen, clean or in use, was sent to the laundry. Admissions to the ward had been severely restricted for 12 weeks and it was closed for a fortnight.

The failure to stamp out the outbreak when the ward was first closed may have been because the search for mild and asymptomatic cases among patients and staff was not sufficiently vigorous.

Medical devices made of plastic

The first exception involves medical devices made of plastic. *Staph. epidermidis* adheres strongly to the surface of plastic and if the surrounding environment is nutritious, they multiply in the biofilm that develops on its surface. This association became obvious when shunts made of plastic were inserted to relieve hydrocephalus in infants with spina bifida. These were almost always colonized, antimicrobial treatment was nearly always ineffective and most of them had to be removed. This pattern has been repeated with the catheters used for continuous ambulatory peritoneal dialysis, with peritonitis rather than septicaemia as the result. These colonizations and infections are nearly all self-infections, perhaps of the autogenous type, and may involve multiply resistant bacteria. They are difficult to control.

Eye infections

The second exception concerns infections of the eye. The inside of the eye is well protected from microbes, but if they do gain entry it is poorly equipped to deal with them. If the endophthalmitis that results is not energetically treated the eye is usually destroyed. *Staph. aureus* and particularly *Pseudomonas aeruginosa* are important pathogens in this situation, the latter perhaps originating in unsterile equipment or medicaments. Postoperative endophthalmitis is fortunately rare, but a perforated corneal ulcer can lead to it. *Ps. aeruginosa* and other pseudomonads may survive and even grow in solutions of the weak disinfectants used to preserve eye drops. It is possible, inadvertently, to treat a patient with drops that contain a major pathogen for the eye, with potentially disastrous consequences. This is why eye drops must be sterilized carefully in small quantities for use on one occasion only or for repeated use over a short period in the same patient. It is dangerous to use multidose containers of eye drops for several patients for periods of days or weeks. This also applies to the solutions used by those who wear contact lenses.

Several diagnostic and therapeutic instruments used in ophthalmological practice come into contact with the surface of the eye. Between uses such apparatus, or at least those parts of it that make the contact, must be sterilized or at least subjected to a high level of disinfection. Failure to do this has in the past led to outbreaks of IaI. The microbes involved have usually been viruses, adenoviruses in particular. On several occasions inadequate disinfection has allowed tonometers to transfer infections from the eyes of one patient to those subsequently exposed to the same apparatus.

INFECTIONS WITH 'DIFFICULT' MICROBES

Of superman, super-rodents and superbugs: MRSA, VRE, *Clostridium difficile*, *Klebsiella pneumoniae*, *Mycobacterium tuberculosis* and prions.

Introduction

Rats and mice with an increased tolerance to warfarin began to appear soon after this anticoagulant was introduced as a rodenticide. Although rodents do not breed as rapidly as bacteria, they are quite prolific and when 'sensitive' animals were poisoned, they were replaced by the 'resistant' strain. Journalists, possibly assisted by scientists with the glitter of research-grant gold in their eyes and a taste for publicity, splashed the news and 'super-rat' and 'super-mouse' were born. These super-rodents were imbued with the strength, boldness and resourcefulness of superman, superimposed on the dirty, vicious and evil reputation of, in particular, the rat. The public was suitably impressed and not a little alarmed. In fact, super-rodents are no different from any other rodents, save in the 'resistance' to warfarin imposed upon them by human intervention.

Modern media people, who may know the difference between a gerund and a gerundive but cannot distinguish between a virus and a bacterium, have now invented the 'superbug'. Once more there has been inside help, perhaps inadvertent, this time from healthcare professionals. The spectre raised is of legions of potent, vicious, invisible enemies that threaten to decimate the human race in the manner of rabbits attacked by myxomatosis. Once more the public is suitably impressed, but this time not a little frightened as well.

As a media superbug, the Ebola virus causes a disease that in some respects resembles myxomatosis but fortunately is less pathogenic and it lacks a vector as superbly efficient as the rabbit flea. A 'flesh-eating' virus emerged in lurid headlines, some time ago, to disappear again a little later. It was in fact the bacterium *Streptococcus pyogenes*, an old sparring partner that stirs itself occasionally to remind us it still exists and that it has not lost its exceptional pathogenic potential. *Strep. pyogenes* is one of the causes of necrotizing fasciitis, a sometimes rapidly spreading and potentially lethal cellulitis that involves the deep fascia. More recently the meningococcus, *Neisseria meningitidis,* has emerged as a candidate superbug. In some respects meningococcal meningitis has begun to follow the path trodden by poliomyelitis in northern Europe some 60 years ago. Under the influence of social and demographic changes poliomyelitis ceased to be a near-universal asymptomatic infection of infants who were immunized harmlessly under the protection of maternal antibody. In the new circumstances (initially unfettered by vaccines that did not appear until the 1950s) poliomyelitis emerged as a paralysing and potentially fatal disease of now non-immune older children and, eventually, of adults as well. Historically the occasional lethal forays of *Vibrio cholerae* and *Yersinia pestis* into northern Europe would have qualified them as superbugs. The thought of plague still frightens the public, as shown by the international reaction to the supposed presence of pneumonic plague in Surat in India in 1994 (the infection was probably due to something else). Tomorrow superbug status may be awarded to the prions or to the influenza virus if or when it reproduces the lethal character it displayed in 1918 and 1919. In a few months significantly more people died of influenza than had been killed in the whole of the First World War.

Superbugs have now invaded the field of iatrogenic infections. The first (though not apparently named as such at the time) was undoubtedly the 'hospital' *Staphylococcus aureus* (HSA) that emerged in the 1940s (Chapter 2). At first this was only distinguishable from any other *Staph. aureus* by its resistance to penicillin and, although the original strains were later discovered to have belonged to a single 'phage type, the resistance soon spread to many others. *Staph. aureus* is an inhabitant of 25–50% of normal individuals. The penicillin-resistant strains carried by many people today (for the most part as harmless colonists) are residues of this episode. There is no hint of anything 'super' about them. With the wisdom of hindsight it is now clear that hospital staphylococci emerged as a problem because penicillin was grossly misused. Their appearance and spread stirred up an equally gross overreaction among the healthcare workers of the time. HSA was widely perceived to be more pathogenic than its sensitive counterpart, though solid evidence for this does not emerge from the scientific literature of the time. What does emerge is that the overuse of penicillin caused the staphylococcal carriage rate to rise to and beyond its normal upper limit. There were many more staphylococci about so there was a higher than usual incidence of otherwise typical staphylococcal sepsis (see, for example, Blowers *et al.*, 1955).

Microbes that have achieved iatrogenic superbug status more recently are the human immunodeficiency virus (HIV), MRSA, VRE, *Clostridium difficile* and multiply resistant strains of *Klebsiella pneumoniae* and *Mycobacterium tuberculosis*. The hepatitis B virus enjoyed a brief spell as a superbug, though because a vaccine is now available its importance has diminished. Together with the HIV virus, it is discussed in Chapter 11, p.236.

MRSA

Methicillin-resistant *Staph. aureus* (MRSA) is now often resistant to one or more aminoglycosides as well (MARSA). MRSA first emerged soon after methicillin was introduced but was not taken seriously until, in 1976, it added resistance to the aminoglycoside gentamicin to its armoury. Now the only appropriate systemic antimicrobial for the treatment of serious infections due to it was vancomycin, an expensive and initially toxic drug. In some countries (for example in the UK) healthcare workers became alarmed, not only for their patients, but for themselves. Loud cries of 'wolf' were heard and some ambulance crews and others began to behave as if they were dealing with an outbreak of pneumonic plague. Patients found to be infected with MRSA and individuals colonized with it were sometimes (and may still be) treated like lepers in the Middle Ages. Official guidelines were published on the control of MRSA.

Obvious parallels exist between the spread of MRSA and the HSA pandemic of the 1950s. With the latter as a model, a number of questions spring to mind. First, does MRSA spread more easily than what is now regarded as 'ordinarily resistant' *Staph. aureus* (ORSA) and is it more pathogenic? Second, what caused MRSA (and MARSA) to emerge and what encouraged it to spread? Third, what, if any, control measures are cost-effective? It is too late to turn the clock back in those countries where considerable effort has already been expended in pursuit of control? Even here an analysis of what, if anything, has been done wrong might be useful for the future.

To try to answer these questions it is necessary to abandon prejudice and to keep an open mind. The general perception is that some identifiable varieties or strains of MRSA (called EMRSA, epidemic MRSA) spread more readily than other varieties. This statement sometimes appears in the abbreviated form, 'spread more readily', a statement subject to misinterpretation. The pathogenicity of MRSA is also said to vary with the variety or strain of MRSA concerned and a second abbreviated statement, 'EMRSA is more pathogenic', joins the first to enhance the superbug image. There is no good evidence that MRSA is, in general, more pathogenic than ORSA; rather the reverse. The ability to spread and to cause infections is and always has been a variable characteristic among different strains of *Staph. aureus*, so it follows that some strains of ORSA spread better and are more pathogenic than some strains of MRSA and *vice versa*. In other words and in these respects, ORSA and MRSA are biologically indistinguishable.

In two hospitals in which ORSA and MRSA coexisted, surveys showed that MRSA accounted for 23% and 52% of all iatrogenic infections due to *Staph. aureus* (French *et al.*, 1990; Meers and Leong, 1990). MRSA can spread but, according to these figures, no more efficiently than ORSA. In the 1950s in some places HSA came to account for 80% of all staphylococcal sepsis. A subordinate question to ask is, does the presence of MRSA add to the total of the infections caused by *Staph. aureus*? So far there is no evidence that it does. When MRSA appears it seems to displace ORSA from an unchanged total incidence of staphylococcal infections rather than to increase that total. As to pathogenicity, the surveys quoted support the idea that there is no difference between the severity of infections caused by ORSA and MRSA. The answer to the first question, then, seems to be 'no'. In present circumstances in terms of epidemiology and pathogenicity, MRSA behaves in the same way as any other staphylococcus.

There can be no doubt that the emergence and the spread of HSA in the 1950s was due to the way penicillin was used at the time, just as the use of warfarin led to the emergence and spread of warfarin-tolerant rodents. The outstanding power of the selective pressure imposed by penicillin can be seen from Table 7.6. Penicillin resistance in the staphylococcus was due to the production of a remarkably active enzyme. The fact that MRSA emerged in a less dramatic fashion might be accounted for by the lower effectiveness against staphylococci of the antimicrobials that followed penicillin (Table 7.6). It is also probably significant that none of these drugs was misused as was penicillin in the 1950s. For these two reasons the selective pressure for the emergence of MRSA and its later spread was less intense. The resistance mechanism is a partial change in the molecule that is the target for methicillin. The very poor performance of third-generation cephalosporins against staphylococci (Table 7.6) may add significance to the temporal coincidence between the introduction and heavy use of these antimicrobials in Singapore and the emergence and spread of MARSA there (Meers and Leong, 1990).

By comparison, then, with the obvious and powerful effect of penicillin in the selection of HSA, the pressure that led to the initial

Table 7.6 A numerical indication ('power index') of the effectiveness of certain antimicrobials against *Staphylococcus aureus*. The 'power index' results from the division of the amount of drug available in the blood of a patient by the least amount needed to disable the bacterium. (In detail, the division of the serum level of each antimicrobial halfway between one dose and the next at the spacing of a standard schedule of treatment, by the minimum amount required to inhibit the growth of a majority of sensitive strains of staphylococci.) The dates shown are approximate. Flucloxacillin came into use in 1970 but its parent, methicillin, appeared in 1960

Antimicrobial	Date introduced		'Power index'
Against penicillin-sensitive Staph. aureus			
Penicillin	1940		67
Ampicillin	1960	IM	17
		Oral	5
Against penicillin-resistant Staph. aureus			
Flucloxacillin	1960/1970		16
Cephalexin	1967		<1
Cefuroxime	1976		16
Cefotaxime	1978		<1
Ceftazidime	1980		1
Ceftriaxone	1980	bd	8
		od	5
Gentamicin	1964		20

Key: IM, intramuscular; bd, twice a day; od, once a day.
Data are taken from Garrod, Lambert and O'Grady, 1985, Simon, Stille and Wilkinson, 1985 and Kucers and Bennett, 1989.

emergence of MRSA is less obvious. The appearance of MRSA coincided with the introduction of methicillin and the broad-spectrum penicillins and cephalosporins, all of which have a reduced, poor or very poor performance against *Staph. aureus* compared with penicillin itself (Table 7.6). The same factors would then have applied a comparatively low level of pressure that encouraged the newly emergent MRSA to spread (p.61). The situation is somewhat different from that which is assumed to have led to the appearance of HSA.

The answer to the second question, then, is uncertain. A supplementary question is to ask what will happen if, or perhaps when, MRSA adds resistance to vancomycin to its armoury. The superbug or doomsday answer is that those who must treat infections with 'VMRSA' will be faced with the therapeutic impotence of the 1930s, a situation repeated when HSA appeared in the 1950s. This is too pessimistic a view. The extended antibiograms of collections of MRSA (including EMRSA) show that without exception they all have chinks in their armour that will admit one or more of the modern antimicrobial drugs that were not available in the 1950s. It has already been noted that MRSA is not universally resistant to the aminoglycosides and the same applies to the macrolides, quinolones, fusidic acid and rifampicin. Although these are not all first-choice drugs for the treatment of severe infections, this is not therapeutic impotence.

The third question concerns the cost-effectiveness of control measures. Current recommendations include the following.

- Widespread bacteriological screening in the search for cases and carriers.
- The isolation of patients who are or who might be infected.
- The isolation of patients who are or who might be carriers and the exclusion of carriers among the staff, with treatment of both groups in an attempt to get rid of MRSA.
- The closure of wards when judged to be necessary.
- More or less extensive sanitary measures applied to the environment.

Without regard to which if any of these measures are effective (none has been subjected to a proper clinical trial), the fact that in the modern global village they are not applied universally means that they cannot prevent the emergence of VMRSA or, insofar as they may be effective, hope to do more than delay the spread of MRSA in those places where they are applied (Chapter 2). Those who have used financial arguments to support the application of control measures have included in their calculations the very considerable cost of the control measures themselves. When such measures are not applied these costs disappear and this includes those related to the screening for and the treatment of carriers. The residual expense is the cost of the vancomycin used in the treatment and prophylaxis of MRSA infections. Although it is an expensive drug, the relative importance of this is diminished when compared with the total cost of medical treatment. Examined in this way arguments in favour of heavy-handed control measures appear less impressive.

ORSA and MRSA appear to be equivalent in respect of their ability to spread and cause disease. They only differ in the therapy each requires. What is the justification for the application of control measures to one and not the other? At various times over the last half-century the measures now applied for the control of MRSA have been applied in attempts to control ORSA. They were shown to be without effect or inconclusively effective, with different opinions about their usefulness. The statement that HSA 'disappeared' when methicillin was introduced in 1960 is not the whole story. By 1960 penicillin resistance was widespread among a number of different 'phage types of staphylococci and HSA was well established in the community. The 'pandemic' had in fact become the normal state of affairs and the HSA of the 1950s is still with us as today's ORSA. The gross overuse of penicillin ceased, the rate of staphylococcal carriage fell and with it the excess incidence of staphylococcal sepsis. There was now an effective treatment available for the residual cases of serious staphylococcal infection. Healthcare workers turned their attention to other things. For the time being MRSA is moving along a parallel though, in terms of severity, less extreme path. The varieties of EMRSA now run well into double figures, so the pandemic of MRSA infection has become a multiplicity of little outbreaks and MRSA is now also firmly established in the community. HSA and MRSA are no more superbugs than warfarin-tolerant rodents are super-rodents. Why the fuss?

Vancomycin-resistant enterococci (VRE)

Group D streptococci are subdivided by their ability to grow on media that contain high concentrations of salt and bile into salt- and bile-tolerant enterococci (principally *Enterococcus faecalis* and *E. faecium*) and the less tolerant streptococci, notably *Streptococcus bovis*. These bacteria are opportunistic pathogens able to take advantage of structural or functional abnormalities in their human hosts to cause infections but they are of no importance to healthy individuals. As such they have always been significant causes of IaIs and their significance has increased as medical practice has become more daring in the treatment of conditions formerly regarded as terminal.

Enterococci, normal inhabitants of the human and animal bowel, are inherently somewhat resistant to antimicrobial drugs and serious enterococcal infections have always been difficult to treat. This difficulty was exacerbated when new resistances were acquired (for example to ampicillin and gentamicin) so that the therapeutic importance of vancomycin was increased. When the emergence of clinically relevant resistance to vancomycin (and to another glycopeptide, teicoplanin) was first reported in 1986, the news was unwelcome and it became a cause for concern (Gray and Pedler, 1992). If enterococci can learn to circumvent vancomycin, why not *Staph. aureus*? Vancomycin-resistant enterococci (VRE) have now been reported from several parts of the world and they seem to be a growing problem. They remain, however, low-grade opportunistic pathogens, important only in special units.

Although the appearance of VRE as a cause of human infections has resulted from the use of antimicrobial drugs in hospitals and elsewhere, its original development may have more to do with animal husbandry than with human medicine. Small numbers of VRE were detected in the stools of individuals who had not recently taken antibiotics and had not been in hospital. A further search found VRE in sewage and in the faeces of chickens, pigs and a farm dog, though not of sheep or cattle. VRE was recovered from raw chickens on sale to the public (Bates, Jordens and Selkon, 1993). Peptide and other antimicrobials are used to prevent infections and increase the rate of growth among intensively reared farm animals. It seems likely that VRE is now widely distributed, individually in small numbers, among the human population who have derived it from the food they eat. They emerge to cause infections when carriers become ill, are admitted to hospital and are exposed to vancomycin or to another antimicrobial that can exert the necessary selective pressure. Transmission from one person to another is then possible, with outbreaks of VRE cross-infection in units where patients in need of continuous ambulatory peritoneal dialysis (CAPD), for example, are managed. The importance of this is that, in these circumstances, measures to control cross-infection are only partially successful. The widespread presence of small numbers of VRE in the human population means that infections due to them will not go away until the antimicrobial responsible for the amplifying effect is no longer used (Frankel, 1994). The underlying problem relates to the vexed question of intensive animal husbandry and the misuse of antimicrobials that make this possible (Perez-Trallero et al., 1993).

Clostridium difficile

C. difficile has been implicated as the cause of colonic disturbances that range from simple diarrhoea to the potentially lethal pseudomembranous colitis. Cases present individually or as outbreaks in hospitals and nursing homes. The elderly and immunocompromised are at special risk, but children and younger adults may also be involved. There is a strong association with the prior use of antimicrobials, but cases of it were described as complications of abdominal surgery long before these drugs were introduced.

A constant feature seems to be a disturbance of the bowel flora, particularly of its anaerobic component, and the introduction or proliferation of a pre-existing C. difficile, although other microbes have been blamed. C. difficile produces a powerful toxin. The epidemiology of outbreaks of the condition is far from clear. Handborne cross-infections and environmental infections have both been identified as contributory. C. difficile produces spores that survive well in the environment, so soiling by incontinent patients may be important. It has been claimed that patients who subsequently occupy the same bed or room are at increased risk of infection and control has proved difficult. The application of measures against cross-infection (hand hygiene and the wearing of gloves) and sanitary measures applied to the environment have been the principal methods used, with ward closures as a last resort. These attempts at control have not been conspicuously successful. Cartmill et al. (1994) have described a large outbreak of C. difficile infection.

As human and animal bowels are the normal habitat of both enterococci and C. difficile, it is worth considering the possibility that a foodborne route may play a part in the epidemiology of C. difficile diarrhoea, as seems to be the case with VRE (above). This would explain the failure of methods designed to control cross-infection when they are applied to outbreaks of C. difficile diarrhoea. Clabots et *al.* (1992) describe a 10% carriage rate of *C. difficile* among patients newly admitted to hospital. Their data can be interpreted to indicate the widespread, low-level presence of this anaerobe in the general human population. As noted elsewhere, a negative culture does not indicate the absence of carriage, as the result is no more than a reflection of the sensitivity of the test used to detect it.

Klebsiella pneumoniae

For many years *Klebsiella* spp. have been recognized as causes of outbreaks of IaI of the urinary and respiratory tracts and of wounds, with septicaemias secondary to any of these. Klebsiellas (and other enterobacteria) began to prove their ability to develop antimicrobial resistance some time ago (Chapter 2) and this has continued. In the last few years klebsiellas resistant to all the penicillins and aminoglycosides, together with the newer third-generation broad-spectrum cephalosporins, have appeared in various parts of the world. The factor responsible is a plasmid that allows the production of an extended-spectrum beta-lactamase (ESBL). The ESBL *K. pneumoniae* has appeared in units in which the third-generation cephalosporins are heavily used (Wallace *et al.*, 1995). Some of them appear to be sensitive only to imipenem and others, for the time being, are sensitive to some quinolones and ticarcillin-clavulanate as well. As noted elsewhere, Gram-negative rods are prepared to move onto the centre of the IaI stage as the pathogens of most consequence if the staphylococci were to desert it. Their array of resistances greatly complicates the therapy of seriously ill patients in intensive care units.

Mycobacterium tuberculosis

The only sources of the human tubercle bacillus are individuals infected with *M. tuberculosis* in whom the lesions have direct access to the outside world so their bacterial causes can escape to infect others. This happens most commonly in cases of pulmonary tuberculosis that have advanced to the open (infectious) stage. The control of human tuberculosis (HTB) is based on the discovery of such cases and of cases of closed (non-infectious) disease before it has progressed to the open stage. The individuals concerned are treated with multiple antituberculous drugs (p.61), first to render them noninfectious and then to cure them. In advanced disease where there has been extensive destruction of tissue surgical intervention may be required to supplement the normal healing process. Many believe the (still controversial) BCG vaccine contributes usefully to prevention. (The epidemiology and control of HTB due to the less common *M. bovis*, the bovine tubercle bacillus, is different.)

The tubercle bacillus multiplies very slowly and it produces a chronic, progressive infection with no dramatic onset to aid prompt recognition. Despite the difficulty of early diagnosis, the more advanced countries had until recently made great strides towards the final control of HTB. The situation in less developed parts of the world is much less satisfactory, with 7.5 million new cases in 1990, an incidence that is predicted to rise to 10.5 million in 2000, to reach a total of some 90 million new cases in the decade. Deaths from HTB were estimated at 2.5 million in 1990, with a million more each year by the end of the century.

The improvement noted in advanced countries has gone into reverse and the deterioration has been accompanied by the appearance and spread of multiply resistant *M. tuberculosis*. This has been noted in two settings. The first is where treatment is inadequate. This may be due to non-compliance among unsupervised patients who fail to take the drugs provided or to inadequate self-medication in countries where drugs are available to the public without prescription. The second setting is in the context of HIV infection and the management and treatment of patients with AIDS (Enarson *et al.*, 1995).

These developments are of acute concern to healthcare workers. They have a genuine fear, substantiated in practice, that not only can they acquire tuberculosis from a patient but also that their infection may be very difficult to treat. A second concern arises from the fact that tuberculosis has been shown to spread in special units set up for the care of patients with AIDS (Wenger *et al.*, 1995). Protection of the healthcare workers concerned has become an important issue (Wilcox, 1995). The use of BCG vaccine is to be encouraged in these circumstances.

It has been predicted that *M. tuberculosis* will cause 30 million deaths in the 1990s, to include a quarter of those among adults that ought to be preventable (Anon, 1994). This and the emergence of multiply resistant variants means that *M. tuberculosis* is the only microbe among those discussed here that has truly earned superbug status.

Prions (proteinaceous infectious particles)

Prions are the cause of a number of mammalian diseases grouped together as the **spongiform encephalopathies**. Among these, scrapie in sheep has been known for centuries and the bovine form (bovine spongiform encephalopathy, BSE, or mad cow disease) has achieved recent prominence. The human diseases known to be due to prions are Creutzfeld–Jakob disease (CJD), with its familial form, the Gerstmann–Straussler syndrome and kuru, in which transmission resulted from cannibalism. The status of an atypical form of CJD that may have been transmitted by eating cattle infected with BSE remains to be settled.

Prions are unique among microbes as they consist of an aberrant protein, entirely free of nucleic acid, that is capable of replicating itself inside living cells. Prions are highly resistant to most of the usual chemical disinfectants and to heat. They can be inactivated in an autoclave at 134°C for 18 minutes (six times the normal time, but see Taylor, 1996) or 121°C for over an hour or by exposure to hypochlorite at a strength of 5000 ppm available chlorine or to a 5% solution of sodium hydroxide, in each case also for an hour.

As a cause of IaI, transmission of CJD has followed the use of human pituitaries taken post-mortem from infected individuals and made into drugs (human growth hormone and gonadotrophin) and of corneas and portions of dura mater used as transplants. Transmission has also been documented between neurosurgical patients although the equipment thought to have been responsible had been cleaned and subjected to conventional sterilization.

The present outbreak of a CJD-like, BSE-related disease in the UK and elsewhere has implications for the future of disinfecting and sterilizing practice and for other aspects of infection control. If the outbreak continues to develop, a growing number of patients who are incubating the condition, as well as those in its terminal stages, will appear in medical establishments. More needs to be known about the spatial and quantitative distributions of prions in the tissues of those affected and an estimate is required of the infective dose. With this information it will be possible to begin to design sensible countermeasures.

REFERENCES

Alexander, W., Fischer, J. E., Boyajian, M. *et al.* (1983) The influence of hair-removal methods on wound infections. *Archives of Surgery*, **118**, 347–52.

Anon (1974) Microbiological hazards of intravenous infusions (leading article). *Lancet*, **i**, 543–4.

Anon (1991) Catheter-acquired urinary tract infection (leading article). *Lancet*, **338**, 857–8.

Anon (1994) The global challenge of tuberculosis (leading article). *Lancet*, **344**, 277–8.

Bates, J., Jordens, Z. and Selkon, J. B. (1993) Evidence for animal origin of vancomycin-resistant enterococci. *Lancet*, **342**, 490–1.

Blowers, R., Mason, G. A., Wallace, K. R. *et al.* (1955) Control of wound infections in a thoracic surgery unit. *Lancet*, **ii**, 786–94.

Brewer, G. E. (1915) Studies in aseptic technic. *Journal of the American Medical Association*, **64**, 1369–72.

Byrne, D. J., Lynch, W., Napier A. *et al.* (1994) Wound infection rates: the importance of definitions and post discharge wound surveillance. *Journal of Hospital Infection*, **26**, 37–43.

Cadwallader, H. L., Bradley, C. R. and Ayliffe, G. A. J. (1990) Bacterial contamination and frequency of changing ventilator circuitry. *Journal of Hospital Infection*, **15**, 65–72.

Cartmill, T. D. I., Panigrahi, H., Worsley, M. A. *et al.* (1994) Management and control of a large outbreak of diarrhoea due to *Clostridium difficile*. *Journal of Hospital Infection*, **27**, 1–15.

Clabots, C. R., Johnson, S., Olson, M. M. *et al.* (1992) Acquisition of *Clostridium difficile* by hospitalized patients: evidence for colonized new admissions as a source of infection. *Journal of Infectious Diseases*, **166**, 561–7.

Colebrook, L. (1955) Infection acquired in hospital. *Lancet*, **ii**, 885–91.

Colebrook, L., Duncan, J. H. and Ross, W. P. D. (1948) The control of infection in burns. *Lancet*, **i**, 893–9.

Cruse, P. J. E. and Foord, R. (1980) The epidemiology of wound infection. *Surgical Clinics of North America*, **60**, 27–40.

Culver, D. H., Horan, T. C., Gaynes, R. P. *et al.* (1991) Surgical wound infection rates by wound class, operative procedure, and patient risk index. *American Journal of Medicine*, **91** (supplement 3B), 152–7.

Enarson, D. A., Grosset, J., Mwinga, A. *et al.* (1995) The challenge of tuberculosis: statements on global control and prevention. *Lancet*, **346**, 809–19.

Frankel, D. H. (1994) Resistant enterococci in California. *Lancet*, **343**, 1560.

French, G. L., Cheng, A. F. B., Ling, J. M. L. *et al.* (1990) Hong Kong strains of methicillin-resistant and methicillin-sensitive *Staphylococcus aureus* have similar virulence. *Journal of Hospital Infection*, **15**, 117–25.

Garibaldi, R. A. (1992) Catheter-associated urinary tract infection. *Current Opinion in Infectious Diseases*, **5**, 517–23.

Garrod, L. P., Lambert, H. P. and O'Grady, F. (1985) *Antibiotics and Chemotherapy*, 5th edn, Churchill Livingstone, Edinburgh.

Gray, J. W. and Pedler, S. J. (1992) Antibiotic-resistant enterococci. *Journal of Hospital Infection*, **21**, 1–14.

Haley, R. W., Culver, D. H., Morgan, W. M. *et al.* (1985) Identifying patients at high risk of surgical wound infection. *American Journal of Epidemiology*, **121**, 206–15.

Hubble, M. J., Weale, A. E., Perez, J. V. *et al.* (1996) Clothing in laminar-flow operating theatres. *Journal of Hospital Infection*, **32**, 1–7.

Krizek, T. J. and Robson, M. C. (1975) Biology of surgical infection. *Surgical Clinics of North America*, **55**, 1261–7.

Kucers, A. and Bennett, N. McK. (1989) *The Use of Antibiotics*, 4th edn, Heinemann, Oxford.

Lidwell, O. M., Lowbury, E. J. L., Whyte, W. *et al.* (1982) Effect of ultraclean air in operating rooms on deep sepsis in the joint after total hip or knee replacement: a randomised study. *British Medical Journal*, **285**, 10–14.

Lynch, W., Davey P. G., Malek, M. *et al.* (1992) Cost-effectiveness analysis of the use of chlorhexidine detergent in pre-operative whole-body disinfection in wound infection prophylaxis. *Journal of Hospital Infection*, **21**, 179–91.

Maki, D. G. (1994) Yes, Virginia, aseptic technique is very important: maximal barrier precautions during insertion reduce the risk of central venous catheter-related bacteraemia. *Infection Control and Hospital Epidemiology*, **15**, 227–30.

Meers, P. D. (1983a) Ventilation in operating rooms. *British Medical Journal*, **286**, 244–5.

Meers, P. D. (1983b) Ventilation in operating rooms. *British Medical Journal*, **286**, 1215.

Meers, P. D. and Leong, K. Y. (1990) The impact of methicillin- and aminoglycoside-resistant *Staphylococcus aureus* on the pattern of hospital-acquired infection in an acute hospital. *Journal of Hospital Infection*, **16**, 231–9.

Meers, P. D., Ayliffe, G. A. J. Emmerson, A. M. *et al.* (1981) National survey of infections in hospitals, 1980. *Journal of Hospital Infection*, **2** (supplement), 23–8.

Meleney, F. L. (1935) Infection in clean operative wounds. *Surgery, Gynecology and Obstetrics*, **60**, 264–76.

Miles, A. A. (1944) Epidemiology of wound infection. *Lancet*, **i**, 809–14.

Mishriki, S. F., Law, D. J. W. and Jeffrey, P. J. (1990) Factors affecting the incidence of postoperative wound infection. *Journal of Hospital Infection*, **16**, 223–30.

Mulhall, A. B., Chapman, R. G. and Crow, R. A. (1988) Bacteriuria during indwelling urethral catheterisation. *Journal of Hospital Infection*, **11**, 253–62.

National Research Council (1964) National Research Council uV study. *Annals of Surgery*, 160 (supplement), 1–132.

Parras, F., Ena, J., Bouza, E. *et al.* (1994) Impact of an educational programme for the prevention of colonization of intravascular catheters. *Infection Control and Hospital Epidemiology*, **15**, 239–42.

Perez-Trallero, E., Mercedes, U., Lopategui, C. L. *et al.* (1993) Antibiotics in veterinary medicine and public health. *Lancet*, **342**, 1371–2.

Schweizer, R. T. (1976) Mask wiggling as a potential cause of wound contamination. *Lancet*, **ii**, 1129–30.

Seropian, R. and Reynolds, B. M. (1971) Wound infections after preoperative depilatory versus razor preparation. *American Journal of Surgery*, **121**, 251–4.

Simon, C., Stille, W. and Wilkinson, P. J. (1985) *Antibiotic Therapy in Clinical Practice*, Schattauer, Stuttgart.

Stamm, W. E. (1991) Catheter-associated urinary tract infections: epidemiology, pathogenesis, and prevention. *American Journal of Medicine*, **91** (supplement 3B), 65–71.

Stamm, W. E. (1992) Nosocomial urinary tract infections, in *Hospital Infections*, 3rd edn, (eds J. V. Bennett and P. S. Brachman), Little, Brown, Boston, pp. 597–610.

Stickler, D. J. (1990) The role of antiseptics in the management of patients undergoing short-term indwelling bladder catheterisation. *Journal of Hospital Infection*, **16**, 89–108.

Taylor, D. M. (1996) Creutzfeldt–Jakob disease. *Lancet*, **347**, 1333.

Treacy, P. J. and Johnson, A. G. (1995) Is the laparoscopic bubble bursting? *Lancet*, **346** (supplement), 23.

Wallace, M. R., Johnson, A. P., Daniel, M. *et al.* (1995) Sequential emergence of multi-resistant *Klebsiella pneumoniae* in Bahrain. *Journal of Hospital Infection*, **31**, 247–52.

Wenger, P. N., Otten, J., Breeden, A. *et al.* (1995) Control of nosocomial transmission of multidrug-resistant *Mycobacterium tuberculosis* among healthcare workers and HIV-infected patients. *Lancet*, **345**, 235–9.

Wenzel, R. P. (1995) The economics of nosocomial infection. *Journal of Hospital Infection*, **31**, 79–87.

White, J. V. (1991) Laparoscopic cholecystectomy: the evolution of general surgery. *Annals of Internal Medicine*, **115**, 651–3.

Wilcox, M. H. (1995) Protection against hospital-acquired tuberculosis, American style: a report on the 4th annual meeting of the Society for Hospital Epidemiology of America (SHEA), New Orleans, 1994. *Journal of Hospital Infection*, **29**, 165–8.

FURTHER READING

Bennett, J. V. and Brachman, P. S. (1992) *Hospital Infections*, 3rd edn, Little, Brown, Boston.

Catanzaro, A. (1995) Preventing nosocomial transmission of tuberculosis. *Lancet*, **345**, 204–5.

Mandell, G. L., Bennett, J. E. and Dolin, R. (1995) *Principles and Practice of Infectious Diseases*, 4th edn, Churchill Livingstone, New York.

Van Griethuysen, A. J. A., Speis-Van Rooijen, N. H. and Hoogenboom-Verdegaal, A. M. M. (1996) Surveillance of wound infections and a new theatre: unexpected lack of improvement. *Journal of Hospital Infection*, **34**, 99–106.

8 Practices in infection control

CARE OF THE HANDS

The how, why and when of handwashing and some alternative strategies. Handwashing is regarded as central to the control of infection. It doesn't happen as often as it should.

Introduction

Everyone knows they should wash their hands after touching a patient, before going on to the next. Observation shows that this happens less often than it should, an impression that is backed up by measurement. In a survey in an intensive care unit (ICU) doctors were observed to wash their hands only 28 times out of each 100 occasions on which they should have done so. Comparable figures for nurses, radiographers and respiratory therapists were 43, 44 and 76 respectively. During the period of observation hands were washed 494 times (41%) out of an ideal 1212 occasions. Because they touch patients more often than anyone else, nurses contributed 400 to the total of 718 failures to wash (Albert and Condie, 1981). Nurses restrict their contacts to a smaller number of patients, however, so the result of their failure was at least limited in its geographical extent. Doctors tend to a larger number of usually more scattered patients so their failure, geographically as well as in absolute numbers, was more serious. In another study a comparison was made between practices in an ICU before and after an upgrading that included an eightfold increase in the number of sinks for handwashing. The frequency of handwashing was a little higher in the new unit, rising to just 30% of the expected figure (Preston, Larson and Stamm, 1981)! More recent observation (most effectively as a patient) suggests that the situation has not improved (see also Zimakoff, Stormark and Olsen Larsen, 1993).

The purpose of handwashing is to remove dirt and reduce the load of bacteria on the skin of the hands (for reviews see Reybrouck, 1986; Ayliffe, 1992). A wash with ordinary soap and water removes much of the transient bacterial flora. This outcome is improved if an antiseptic soap (with chlorhexidine or povidone iodine) is substituted for ordinary soap (p.123). In hospitals this often includes important potential causes of handborne cross-infection (Maki, 1986; Sanderson and Weissler, 1992; p.48). The resident flora of healthy hands is less likely to cause infections, though this does not apply to immunocompromised patients. Resident bacteria are reduced in number by handwashing, particularly if a detergent containing an antiseptic is used, but they cannot be eliminated (Jarvis *et al.*, 1979). Among these residents *Staphylococcus epidermidis* is now an important pathogen for immunocompromised patients. Unlike *Staph. aureus* with a carriage rate of 25–50%, *Staph. epidermidis* is a member of the resident flora of 100% of the population. Among a group of healthcare workers in an ICU, 80% were found to carry the methicillin-resistant variant of *Staph. epidermidis* on their hands (Maki, 1986).

Trivial injury profoundly alters the resident flora of the skin. Damage may be due to the minor traumas hands suffer every day or be caused by failure to rinse or dry the hands properly after washing, by the frequent use of strong detergents and antiseptics or by the vigorous use of a nail brush (Ojajarvi, Makela and Rantasalo, 1977). When the hands of healthcare staff are damaged the new and more numerous bacterial residents may include important causes of iatrogenic infections, including MRSA. These have become true residents so are not removed by washing, even if an antiseptic is used. Staff in this condition are a serious threat to patients. Because the injury may be inconspicuous (a tiny hang-nail, for example) an individual may ignore or even be unaware of it. Unless such people are tested bacteriologically they cannot know they are carrying important pathogens (Meers and Leong, 1990). It is clear that the simple injunction 'now wash your hands' is a long way from providing the final answer to handborne cross-infection.

Alternatives to handwashing

Alcohol (spirit) is a powerful antiseptic that evaporates quickly (p.122). Applied to the skin at about 70% in water, it rapidly kills transient and a proportion of the resident bacteria and then disappears. When it is desired to reduce the bacterial load on hands that are not obviously soiled, washing can be replaced by the application of alcohol. The inclusion of a non-volatile antiseptic (chlorhexidine p.123) adds a residual effect to the immediate action of the alcohol. If an emollient like glycerine and perhaps a perfume are added the result is a degerming lotion or handrub that does not dry the skin too much (Kurtz and Boxall, 1976). A number of commercial products meet this specification. The wearing of gloves offers a third, most effective, but more expensive way of preventing the transfer of organisms to and from the hands. There has been a recent significant increase in the use of gloves throughout the healthcare industry. This is a reaction to the perceived risk of occupational HIV infection, so the motivation is reversed; for the benefit of the healthcare worker rather than the patient. This change of emphasis does not seem to have made any difference to handwashing practice, indeed, probably the reverse. If the same gloves are worn to complete multiple separate procedures they must be washed in between in exactly the same way as the naked hands (p.196).

High-risk areas

A dilemma arises in ICUs and in other areas where immunocompromised patients are cared for, where the risk of IaI is high and handwashing ought to be more frequent. The use of antiseptic detergents that incorporate chlorhexidine or an iodophor is often encouraged in such places with the sensible intention of reducing the bacterial population of the hands. However, if this leads to damage to the skin then the hands may be more dangerous than if ordinary soap had been used or even if they had not been washed at all. A way out of the dilemma might be to provide several different handcare regimes in high-risk areas, to be used alternatively or sequentially at the discretion of members of staff or according to some protocol. The object would be to reduce the possibility of damage to the hands that can follow the repeated use of a single method on its own. Such techniques as washing with ordinary soap, with antiseptic detergents (perhaps rotating between different products), the use of a degerming lotion, the wearing of gloves and the washing or antiseptic degerming of gloved hands between procedures might be used sequentially or as the task to be performed dictates. In other clinical areas handwashing with ordinary soap is generally thought to be sufficient, though degerming lotions are useful as a quick alternative during procedures or in the course of duties that demand more than one handwash to complete them. Degerming lotions have a particular application where facilities for washing hands are inadequate.

How and for how long?

Nail brushes should be used very little if at all and only to remove dirt that is ingrained or trapped under the nails and for the latter a scraper may be substituted. A nail brush plus a strong detergent is much more likely to damage the skin. Microbiologically, the traditional form of 'scrubbing up' in surgery for five to ten minutes does not seem to improve on an energetic wash with an antiseptic detergent for two or three minutes. Handwashing between cases on a surgical list may be restricted to one minute. Outside operating departments a 15-second wash probably suffices for hands that are already 'clean', though somewhat longer is required for hands that are 'dirty' (badly contaminated). A point worth remembering is that the water used for washing hands is rarely anything like sterile; indeed, it may contain surprisingly large numbers of bacteria. This is one of the reasons why a rub with alcohol (with or without added residual antiseptic) improves the hygiene of the hands after a surgical 'scrub'.

The technique of washing requires attention. Experiments have shown that many people fail to wash their thumbs properly when they wash their hands. Wet hands are socially unacceptable and careful atraumatic drying also serves to lessen the damage that accompanies frequent washing. Disposable paper towels or fabric towels in rolls inside a machine that offers a fresh, dry portion to each user are commonly employed. Both require a foolproof system to ensure constant availability. Hand hygiene collapses entirely when confronted by an empty paper towel dispenser or the tail end of a fabric towel that has clearly already been used by many others. Paper towels should be soft to the touch. Harsh towels damage the hands and staff may not use them. Fibres rubbed from paper towels are a component of the dust found in places where they are used.

Conclusion

Hands are inseparable from healthcare. Hands are also recognized to be important causes of IaI. There is a general consensus that a significant part of this might be prevented by proper care, generally construed as more and better handwashing. Yet in practice it seems that hands are washed less than half as often as they should be. As this failure to wash was measured in ICUs where the importance of handwashing is likely to have been stressed, the situation is probably worse elsewhere in hospitals and in other medical establishments. We do not know if this represents a lack of education or a stubborn refusal to accept unwanted and perhaps impractical advice. Attitudes and motives are important (see, for example, Larson and Killian, 1982; Williams and Buckles, 1988; Box 8.1). There is a need for more effective and less damaging handcare methods. Hot air hand-dryers would lessen the damage done to hands by disposable towels made of the rougher kinds of paper. Their use would certainly reduce the volume of hospital waste and perhaps reduce costs as well. Some

---

Box 8.1   Why don't I wash my hands?

Lack of leadership
- Others (particularly senior nurses and doctors) don't wash their hands, so why should I?

Lack of time, low priority
- I'm too busy, other things are more important (my hands are not *really* dangerous, are they?).

Lack of facilities
- Sinks are too few or inconveniently sited; I don't like the soap or detergent supplied or none is available; the method provided for hand-drying is inadequate or is absent.

Lack of motivation
- Frequent washing of my hands makes them sore.

Lack of memory
- Well, I forgot!

newer models are less noisy than their predecessors, though they still take what many regard as an unacceptable time to dry the hands. This might deter even more people from washing them in the first place. However, it should be noted that there are no microbiological contraindications to their use in clinical areas (Mathews and Newsom, 1987; Meers and Leong, 1989).

CATHETERS AND OTHER TUBES

Diagnostic and therapeutic tubes and other invasive devices all have much in common, including the fact that, potentially, they are killers. The application of well-thought out infection control strategies can reduce the risk. Epidural catheters are discussed.

Introduction

Intravascular and urinary catheters, endotracheal and peritoneal dialysis tubes, the tubes used for epidural anaesthesia and for drainage in surgery, together with other invasive devices, have one major feature in common. They all open highways through a body's normal defences along which microbes reach, colonize and infect places otherwise inaccessible to them. Hollow tubes give microbes a choice of routes by which to travel, either round the outside (**extraluminal spread**) or through the lumen of the tube itself (**intraluminal spread**). Infections that result from extraluminal spread are almost entirely endogenous (or autogenous) self-infections and they arise from the normally or abnormally colonized skin or mucous membrane through which the tube enters the 'forbidden zone' (Figure 3.1; p.19). Infections following intraluminal spread are usually exogenous, the microbes concerned arriving as a result of cross- or environmental infection (p.48).

Intra- and extraluminal spread

Depending on circumstances, the extraluminal (endogenous or autogenous) route of infection may be more or less common than the intraluminal (exogenous) route. There are a number of important variables. These are:

- whether the tube is filled with liquid or air;
- the direction of flow;
- the weight of colonization of the body surface at the point of entry or exit of the tube;
- the level of care exercised in the insertion and management of the system.

Colonization and infection of that part of the forbidden zone entered by the tube (p.19) becomes increasingly common as time passes. To slow this down or prevent it, an assessment needs to be made to determine which route of infection is more important. Control measures designed to delay the appearance of infection vary according to the route followed and it is the job of the infection control team (ICT) to recommend appropriate procedures to reduce the risk. Most tubes are left in place for a very few days. In these cases management that delays the entry of microbes by the most important route will be cost-effective if the tube has been removed by the time infection would otherwise become a serious threat. It is important for staff to understand how infections develop so that they know what is happening and do not become frustrated by apparent failure and stop trying to control them.

Control of the **extraluminal routes** has been discussed in Chapter 7. The methods can be summarized as choice of tube, care at insertion, the use of antiseptics and scrupulous maintenance. The materials used in the construction of the tube (Jansen and Peters, 1991) and its diameter and length are important. As far as the materials are concerned, teflon and silicone, for example, are improvements on polyvinyl chloride (but are more expensive) and a smaller diameter is preferred to a larger one. Insertion should be made with the least possible trauma, to limit damage to the body's innate defences. Where possible the body surface through which the tube is to be introduced should be carefully cleaned and disinfected first. This is to reduce the number of bacteria carried on the outside of the tube into the forbidden zone, as it is inserted. For skin disinfection quick-acting alcoholic solutions are preferred. On mucous membranes less effective and more slowly acting aqueous solutions have to be used. Compared to the near-instantaneous effect of alcohol, these take some minutes to produce the desired effect.

It is not clear if regular toilet of the skin round insertion sites is of use or if antiseptic ointments should be applied. Fiddling with the entry point increases the risk of infection with urinary catheters, so in this case meatal toilet as a special procedure is not encouraged. Washing to a good standard of personal hygiene is sufficient. By extension, these considerations may apply to tubes of other kinds. Although the application of antiseptic ointments is controversial (Chapter 7), topical ointments containing antimicrobial drugs should never be used as they encourage the appearance of resistant strains of bacteria.

Maintenance may be summarized as keeping closed systems closed as much as possible, the use of careful, clean techniques when it is necessary to break into closed systems and adherence to the best advice on the frequency of changing tubes and their associated external apparatus. The latter is doubly important as significant sums of money can be saved by not changing complex and expensive apparatus as often as has sometimes been advocated in the past (Chapter 7).

Intraluminal spread varies in importance. The application of inadequate procedures for the care of urinary catheters is accompanied by unacceptably early colonization and infection due to contamination by the intraluminal route. It is silly to buy modern apparatus for the collection of urine by the closed method and then use it in such a way as to make it open. Intraluminal spread is usually less of a problem with tubes that contain therapeutic liquids, because more attention is paid to these. From the point of view of preventing infection, a urine-collecting system deserves the same level of care as an intravenous drip.

The rate of infection associated with the use of invasive devices is directly related to time for which they are in place. The interval between the insertion of a device and the development of an infection due to it can be increased by the application of sensible control measures. If a device is no longer needed and is taken out before an infection has time to appear, the patient concerned is spared an IaI. Once it has been inserted it should be reviewed regularly to see if it can be removed. It cannot be stressed too much that these devices are potential killers. They should not be used for convenience.

Tubes and other devices of the kinds under consideration have a potential for doing good and harm. To some extent the harm (in this context, the development of an infection) is unavoidable and has to be accepted as the price of doing good. Sensible, well-thought out procedures that are understood and followed by everyone can keep this harm to a minimum. In consultation with those who are to apply them, the ICT has a major role in the development of these procedures, in the education of the staff who must apply them and then in their evaluation (audit, p.66).

Epidural catheters

In recent years there has been a considerable growth in the popularity of the epidural route for anaesthesia and analgesia. Catheters used to introduce the anaesthetic for the induction of medium to long-term analgesia share the characteristics of the tubes discussed above, including a tendency to cause infections. Epidural abscesses and meningitis are the serious complications of the extraluminal spread of infection from skin puncture sites. The incidence of infection has been shown to be directly related to the debility of the patient and, once more, to duration of catheterization (Holt *et al.*, 1995). Epidural catheterization should be performed with the same attention to detail as central venous catheterization.

DRESSING WOUNDS

Money is saved when practices used for the dressing of wounds are simplified. In some circumstances it is possible to make changes without doing any harm.

Introduction

For a complete review of this subject this section should be read in conjunction with Chapter 7, p.135, and Chapter 10, pp.215 and 224. Once more it is necessary to stress the importance of the distinction between wounds that are 'closed' and those that are 'open'. Most ordinary surgical wounds fall into the 'closed' category. Open wounds include surgical incisions that were originally closed but where the suture line has broken down or where primary closure was undesirable or impossible, pressure sores, ulcers of different kinds and burns. There are fundamental differences in the potential for bacterial colonization and infection between the two types of wound, in the rate at which they heal and how this happens. The care required by each is also different. In Chapter 2 and Chapter 7, p.135, the introduction of the 'no-touch' dressing technique was related to problems with infections of wartime wounds. These were, in the main, open wounds. The question is, how much of a technique invented for the dressing of open wounds should be applied to what, in ordinary circumstances, are the much more numerous closed variety? Large sums of money can be saved when simplified methods are used in the care of the latter.

Closed wounds

Most operative wounds are closed completely at the end of surgery, with or without a vacuum drain. If such a wound breaks down later the cause was nearly always present within it at the time the patient left the operating room. The underlying cause is usually surgical (p.138). If the soil (the tissues in proximity to a wound) has inadvertently been fertilized (made more susceptible, Table 7.1), a relatively small and otherwise innocuous dose of bacterial seed planted at the same time can start an infection. In some cases the bacterial seed travels through the blood to settle in the fertilized soil a little later. In these haematogenous infections the source of the bacteria may be a septic condition elsewhere in the body, for example a urinary tract infection.

After a few hours wounds that have been completely closed are effectively sealed to the entry of bacteria from the outside (for more details see Chapter 7, p.141). Because subcutaneous tissues are not exposed there is none of the desiccation that contributes to the delayed healing of open wounds. In these circumstances what happens in the ward postoperatively has little bearing on whether or not they become infected or the rate at which they heal. They certainly do not need to be dressed using an elaborate and expensive so-called aseptic ('sterile') ritual. If anything needs to be done, a simple clean procedure is all that is required.

Open wounds

Wounds in which subcutaneous tissues are exposed without a covering of skin are completely different. Bacterial colonization or infection is inevitable and healing, perhaps already slowed down by a poor blood supply, may be compromised further by desiccation. The deleterious effect of the latter has attracted attention and the correction of it by the use of appropriate types of dressing has improved the outlook for the healing of these wounds. This is of interest microbiologically, because a wound dressed so that it is kept moist provides an environment in which bacteria may grow more freely. Some years ago open wounds and in particular burns, were treated by exposure expressly to keep them as dry as possible.

In many cases the first bacteria to colonize an open wound are derived from the patient's own microbial flora. A hospital strain of bacteria may be involved if the patient's normal flora has been altered because they have already spent some time in hospital. The patient may be nursed with others who have similar wounds. In such cases it has proved difficult or impossible to prevent any 'personal' bacteria in an initial colonization from being replaced by the multiresistant hospital strains that typically infest such clinical areas. Ultimately all open wounds in the same clinical area are colonized or infected by the same bacteria. This process is at its most highly developed in burns units. What determines the point at which a colonization becomes an infection in an open wound is not clear. Stratagems for diagnosing and treating these infections are complex and best determined in the context of local conditions (see Chapter 10, p.224).

Dressing wounds

Changing the dressings on open wounds requires the use of a refined, 'no-touch' technique because the exudate probably contains large numbers of bacteria, which may be of undesirable multiresistant hospital strains. When wounds of both the open and closed varieties are present in the same clinical area, it makes microbiological sense to care for the open kind after any closed ones have been dealt with. This separation is a good idea because the various techniques used to prevent the transfer of bacteria between patients (no-touch, gloves, handwashing) are less than perfect. Even if handwashing fails or is incomplete or forgotten the error is more likely to be forgiven if the hands have not yet been contaminated by discharges from infected wounds. The purpose is to try to prevent the spread of hospital strains of bacteria to patients or staff who are still free of them.

The same logic applies to the dressing of open wounds. If possible they should be dressed in the following order: first, wounds with corrugated or Yeates drains not known to be infected or colonized; then wounds colonized or infected with bacteria that are not hospital strains; and finally, wounds that are already colonized or infected with a hospital strain of bacteria. Of course if there are enough nurses and ward routine allows, a different nurse can attend to each class of wound or one can attend to the open variety while another is dealing with closed wounds.

To avoid the waste inherent in using a classic 'no-touch' technique for all wounds, a decision has to be made in each case. There is no problem with wounds known to be open. Closed wounds are assessed by eye if they are visible or are assumed to be closed if the dressing that covers them is dry. If the dressing is wet the wound is assumed to be infected and a procedure appropriate to an open wound is employed. Dressing may be removed wearing a clean plastic bag on the dominant hand and the bag is turned inside-out to enclose it, for discard. Any other attention thought necessary is performed according to clinical assessment.

It is appropriate to use either gloves or forceps in a sterile procedure. If disposable gloves are chosen they need not have been sterilized and there is no reason to buy an expensive variety. Cheap plastic gloves intended for clinical use from a responsible supplier are virtually free of microbes when taken from their packs, provided the latter are of a sensible design. They should emerge through a small opening in their box and be packed within it so that each protects the one below. Nurses changing dressings should wear plastic aprons to protect their clothing from splashes. These need not be changed between patients when dealing with closed wounds, but they should be changed between dressing open ones. If this is inappropriate due to cost or the fear of producing too much plastic waste, a fresh clean (not necessarily sterile) fabric gown may be used for each patient, worn over a plastic apron that is not changed. However, it is important to recognize that gowns made of ordinary fabrics cannot prevent bacteria contained in splashes or carried on skin squames from passing through them (p.198). Without a plastic apron underneath, they offer no useful protection to either the wearer or the wound. The plastic apron is likely to be contaminated if the gown over it gets wet, so this alternative is less attractive than a plastic apron that is changed between patients with open wounds. Of course, nurses should wash their hands between each procedure. Masks are unnecessary.

A good deal of folklore attaches to the possibility of bacteria being carried from patient to patient on nurses' clothing. Clean clothing that has been properly laundered (p.246) is virtually sterile. If it remains dry any of the small number of bacteria that adhere to it during wear do not survive well. In the absence of major soiling with faeces, heavily infected urine or frank pus, clothing will always have many fewer bacteria on it than are present on the skin of the wearer under it. The transfer of bacteria on a nurse's clothing from one wound to another requires that direct physical contacts are made between them. Bacteria cannot jump, nor are they easily detached from surfaces once they have arrived there, so for practical purposes spread on clothing is not a problem.

Dressings fulfil a fivefold purpose. First, they should protect a wound from infection from the outside for the few hours to a number of days (depending on the type) during which this remains a possibility. Second, in the case of open wounds, they should produce the physical conditions under them required to promote healing. Third, they should soak up excess discharge from the wound. Fourth, they may act to prevent stitches or clips catching on clothing, etc. and fifth, they keep prying fingers at bay.

The importance or otherwise of dressings on closed surgical wounds was highlighted a long time ago by a report of a trial in which surgical wounds of this type were left undressed, without harm to patients (Howells and Young, 1966). It is probable that, after the first few hours, the presence of a dressing on a closed wound fulfils a psychological rather than any real need. A simple clean procedure is used to remove stitches or clips from a closed, healing wound.

At one time it was proposed that each hospital be provided with central treatment areas where all wounds would be dressed. The idea seems to have originated in a proposal made in the restricted circumstances of burns units, in which the infection of wounds by airborne spread of microbes is a possibility (p.224). To extend this to cover the dressing of all wounds, so that much larger numbers of patients from all parts of a hospital are brought to a central facility, there to be exposed to a significantly greater risk of cross-infection, is not recommended.

ISOLATION

The use of isolation in the control of infection and the manner of applying it have been described as a conundrum. The reason for this is discussed and an attempt is made to separate fact from fiction.

Background

The human race usually overreacts to dangers it does not comprehend. The application of isolation to the control of infection is a good example. In the Middle Ages lepers and sufferers from bubonic plague were isolated from the community. This may have done something for peace of mind but it did little or nothing to prevent the spread of either disease. The segregation of people to stop the spread of contagion and the enforcement of quarantine evolved as described in Chapter 2 and Chapter 5, p.82. Although Semmelweis and Lister dealt with specific and classic iatrogenic infections, at the time more attention was paid to the control of such community infections (CIs) as cholera and typhoid. It was not until the 1940s that an integrated approach to the problem of IaI began to emerge. At that time 'barrier nursing' was one of the methods used to prevent the spread of CIs when cases of them were admitted to fever hospitals. As these hospitals were closed (p.82) many of their senior nurses were trans-ferred to work in general hospitals. With them went various techni-ques of barrier nursing which, renamed isolation, were adapted to the control of IaIs.

The need for standardization soon became apparent and in 1970 in the USA the Centers for Disease Control published a handbook on isolation techniques. This was revised in 1975. These publications described various categories of isolation appropriate to the supposed route of transmission of the infections dealt with. During the 1970s attention was increasingly focused on the prevention of HAIs and more had been learnt about their epidemiology. It became apparent that the requirements for isolation were too rigorous and in 1983 and again in 1986, new recommendations were issued (Garner and Simmons, 1983, 1986). These dealt with HAIs more fully, but the recommendations still concentrated on such diseases as tuberculosis, hepatitis B and varicella-zoster. By then these had become minor components of HAI in many of the more developed parts of the world, though this was and is not true everywhere and of course tuberculosis has now re-emerged as a problem in hospitals.

The isolation of patients with infections (**source isolation**) developed in two different ways. In 1983 the CDC described seven different varieties of **category-specific isolation**, designed to deal with the seven routes by which the infections to be isolated were thought to spread. The categories were strict, contact, respiratory, tuberculosis, enteric, drainage–secretion and blood and body fluids (Table 8.1). A separate list of procedures was laid down to counter spread by each of the routes or for certain types of infections. Many hospitals developed broadly equivalent systems, but limited the number of categories to three or four. The precautions in each category had been designed to meet worst-case conditions, so for the average situation they were too demanding. The need for the full list of recommendations was not accepted everywhere and the diseases for which isolation was suggested included a number not thought to require any special precaution in much of the rest of the world.

In the second form of source isolation the precautions were tailored individually to each infection and the process was called **disease-specific isolation**. This was designed to accord with the perceived 'infectiousness' of each type of infection and the route by which it was thought to spread. It was more flexible and less wasteful, but included over 160 different diseases or conditions, each with its own set of precautions (Garner and Simmons, 1986). Its application required a degree of specialist knowledge and skill that may have prevented cate-gorization by ward staff, so some delay was introduced by the need to consult members of the infection control team (ICT).

In 1987 a development and expansion of the former blood and body fluids category of precaution was proposed. This soon came to be called **body substance isolation** (BSI) or **universal precautions** (UP) (Lynch *et al.*, 1987; Centers for Disease Control, 1988). A major

Table 8.1 An outline of the 1983 precautions recommended by CDC for category-specific isolation (condensed from Garner and Simmons, 1983)

Isolation category	Types of infection
A Strict isolation	Disseminated varicella-zoster, pneumonic plague: for Lassa fever, etc. change 2* in Table below to 3.
B Contact isolation	Paediatric patients with acute respiratory infections; neonates with gonococcal conjunctivitis, herpes simplex or staphylococcal skin infection; any patient with cutaneous diphtheria, disseminated herpes, infection or colonization by epidemiologically significant multiply resistant bacteria, staphylococcal pneumonia or major skin infections or infestations, rubella, vaccinia
C Respiratory isolation	Measles, erythema infectiosum, respiratory infection including meningitis due to *Haemophilus influenzae*, invasive meningococcal disease, mumps, pertussis
D Tuberculosis isolation	Actual or suspected open (infectious) tuberculosis
E Enteric precautions	Infectious gastrointestinal disease, including hepatitis A
F Drainage or secretion precautions	Any disease producing an infective purulent discharge, unless already in a more demanding category
G Blood or body fluid	Human immunodeficiency virus infection, arthropod-borne viral fevers, hepatitis B, non-A non-B, leptospirosis, malaria, etc.

	Isolation categories						
Requirements	A	B	C	D	E	F	G
Separate room	2*	2	2	3	1	0	1
Gowns	2	1	0	1	1	1	1
Gloves	2	1	0	0	1	1	1
Masks	2	1	1	4	0	0	0
Handwashing	2	1	1	1	1	1	1
Waste and equipment	6	6	6	5	6	6	6

Key: 0, Not necessary; 1, optional, or only in connection with direct patient contact; 2, necessary; 3, room with special ventilation; 4, masks worn if patient coughing; 5, waste and other fomites treated as mildly infective; 6, as for 5, but special decontamination for reusable items.

stimulus for this was anxiety among healthcare workers about the infections they might acquire occupationally from undiagnosed cases of, in particular, hepatitis B and HIV infections among the patients they cared for. The technique emphasized the application of relatively simple barrier precautions (wearing gloves, for example) for every patient when contact is made with potentially infected (so potentially all) body fluids, secretions and excretions. The need for true isolation in a single room was reduced to a short-list of what for most hospitals were less common conditions. This approach was not accepted everywhere and some hospitals developed a two-tier system that combined UP or BSI with a variable residue of earlier practices.

This two-tier approach received the CDC seal of approval at the beginning of 1996 with the issue of new guidelines (Garner, 1996). These carry forward the ideas central to UP and BSI and include regulations imposed by the US Occupational Safety and Health Administration. They are now called **standard precautions**, to be applied in conjunction with a second tier of **transmission precautions**. The latter reduce the earlier (1983) seven categories of isolation to three (airborne, droplet, blood and body fluids) and add a curtailed list of specific diseases.

A synopsis of the proposals for standard precautions is given in Table 8.2. The transmission precautions are too complex to be summarized usefully; they need to be studied in their entirety. The list of specific diseases provided is conditional. It may be thought that some of the infections named are wrongly categorized and that an element of overinsurance has been perpetuated. The precautions themselves contain let-out clauses. Two categories, airborne transmission and droplet spread, form a continuum that cannot be divided satisfactorily into two components. The document is generally permissive in its tone.

Transfers of patients and staff make it desirable, indeed necessary, for hospitals in the same region or country to develop isolation policies that conform to a central core of practice, though most find it difficult to adopt complete, ready-made packages. In any case the pursuit of comprehension and compliance suggests that each hospital

Table 8.2 An outline of the 1996 CDC guidelines for 'standard precautions', to be observed in the care of all patients (condensed from Garner, 1996)

Personal contacts
1. Applied to all contacts with patients except as modified under 3 or 4
- Wash hands with plain soap, whether gloves are worn or not, or use an antiseptic preparation if specified by the ICT.
2. Applied to contacts with blood or any body fluid, secretion or excretion (except sweat), with broken skin, a mucous membrane or items contaminated by them
- Wear clean, non-sterile gloves.
3. Additionally applied if any of the materials specified in 2 are likely to be sprayed or splashed
- Wear masks and eye protection or a face shield.
- Wear a clean, non-sterile gown.
4. At all times
- Take special care with used sharp objects and instruments, as provided for by the ICT or occupational health department.

The environment, etc.
- Procedures as directed by the ICT.

Key: ICT, infection control team.

or hospital group should study and as necessary modify a more generally agreed core of practice to produce its own set of rules, according to their circumstances. The development of an isolation policy is an important task for individual infection control organizations.

Back to basics

It is sometimes thought necessary to care for patients under conditions of physical isolation. This may be total or apply only to one or more kinds of interpersonal contact. A patient with an infection may be isolated from other patients and from staff, who otherwise might be infected themselves or carry the microbe concerned to infect another patient. This is **source isolation**, equivalent to the old barrier nursing. Alternatively a specially susceptible patient may be isolated to protect them from microbes or an infection carried by others (**protective isolation**, originally called reverse barrier nursing).

To be successful, source isolation depends on the person (or persons) isolated being the only source of the unwanted pathogenic microbe in the hospital or community concerned. In many countries this may be the situation with rare infections such as Lassa fever or diphtheria. It is clearly not true of most cases of IaI in which the pathogens concerned are microbes widely distributed as parts of peoples' normal floras (among carriers) or in established cases of clinical or subclinical infections. Nothing is to be gained by the rigorous isolation of cases of infection with ordinary strains of *Staphylococcus aureus* or *Escherichia coli*. The same thought might be extended to patients who are carriers of or who are infected by, for example, MRSA, unless it is certain that all the others who carry the same organism (patients and staff) are isolated or excluded simultaneously (p.166). This view has sometimes been modified by the thought that isolation might be of limited value if the person isolated is an unusually prolific source of the pathogen concerned. In general, however, unless a complete barrier is erected between all who carry the microbe to be excluded and those who do not, the exercise tends to be an expensive and, in the end, fruitless ritual. A major weakness of isolation (for patients) or exclusion (for staff) is that in most cases neither can begin until a microbiological examination has shown the need for it. This imposes an inevitable delay of 24–48 hours during which time the pathogen to be excluded is likely to spread to others. If spread within such a period is not common, what is isolation for?

Facilities and practices

The features of complete isolation have included the use of single rooms with an airlock anteroom containing a handbasin for staff use, provided with a toilet bay containing another handbasin, water closet and shower for the patient. For source isolation the air supply may be at slightly reduced pressure and the stale air is filtered or is discharged into the open well away from people or air inlets. For protective isolation the air pressure in the room may be kept a little above atmospheric. Staff should of course wash their hands on entering and leaving and, depending on local rules, may be required to don gowns, aprons, masks or gloves in the anteroom and remove them when they leave. As well as regulating these factors, rules may also be made about who and what can enter or leave the room and in the case of source isolation, how anything that is taken out is to be dealt with.

The most complete form of isolation is provided in specialist infectious disease units. Here even the staff caring for the patient may be isolated from the rest of the world, as used to happen with smallpox. In extreme cases a plastic isolator that completely encloses the patient and their bed may be used. The patient is cared for through armholes onto which are sealed long-sleeved gloves. Food, drugs and other items are passed in and out through an airlock. Fresh air is pumped into the isolation capsule, filtered on its way out in the case of source isolation or on its way in for the protective form.

Protective isolation has in the past been used for patients who are immunosuppressed for transplantation or who for some reason are severely immunocompromised. With growing experience and as immunosuppression has become more sophisticated, major centres have reduced the level of isolation they require for these patients. The earlier extremes such as rooms ventilated with ultra-clean (laminar flow) air, the provision of sterilized food, prohibition of physical contact with others and the extensive use of prophylactic antimicrobials or antiseptics have been abandoned. Even single-room isolation is no longer considered essential. One of the reasons for the general relaxation of precautions is the observation that most infections among these patients are self-infections.

Cohort care in infection control

If a significant number of patients are colonized or infected with a pathogen for which isolation is thought necessary, a system of cohort care can be instituted. All the patients affected are moved to (or are kept within) a discrete clinical area where they are cared for by staff who are restricted to these patients. (This is unlikely to be attainable for all categories of staff.) In these circumstances staff who are themselves carriers of the unwanted pathogen may continue to work rather than being excluded. In theory the cohort system of isolation might be effective if the cohort includes all the individuals who are infected with the pathogen in question, plus any who carry it. To achieve this requires a large-scale, very precise microbiological investigation and a standstill on the work of the unit concerned (or at least its closure to new admissions) while everyone is allocated to the correct category and until the incident is resolved. In the face of an apparently intractable outbreak many infection control organizations feel obliged to institute cohort care, though it is usually applied in a less than total form. The fact that almost half of the outbreaks recorded by Haley *et al.* (1985) had resolved without any human intervention must be taken into account when judgements are made about the usefulness of this manoeuvre.

Summary

It is a fact that though many regard isolation as central to the control of infection, there is no good evidence for its efficacy. This is a pity, because isolation is expensive and disruptive. It is likely to damage patients' mental health and if isolation deprives them of any special monitoring or treatment they require, it may damage their physical health as well (Box 8.2). Over recent years the tendency has been to relax requirements for isolation, a trend that seems likely to continue. This relaxation has not been accompanied by an upsurge of IaI so the indications are that airborne and other environmental sources of infection are of small importance. Ultimately the extent to which single-room isolation is still necessary comes into question. **It may be that the contribution of isolation has been to provide a physical barrier that reminds staff to apply the simple precautions they ought to use for all patient contacts.**

There is an astonishing lack of experimental evidence to support the use of isolation or to define which isolation practices are effective. This accounts for the overinsurance common in its application and makes the design of an isolation policy one of the more difficult tasks that faces infection control organizations. Not least among the problems is the need to identify which members of staff are to initiate

---

Box 8.2 Isolation can be dangerous

An overweight lady aged 46 was admitted to hospital with symptoms of lower bowel obstruction. She was found to have severe diverticular disease and underwent a resection. The operation was technically difficult and postoperatively her wound broke down. She went through a stormy period, part of which she spent in an intensive care unit. She was there at the same time as another patient who had a urinary tract infection with an outbreak strain of a multiply resistant *Klebsiella*. Her wound became infected with this bacterium. When she was sufficiently improved to return to an ordinary ward her original surgical ward refused to accept her as she was still discharging the *Klebsiella* from an abdominal drain. She was transferred to the hospital isolation unit. There she seemed to do well for a few days, but then began to go downhill. By the time it was realized that she had developed a leak in her anastomosis, peritonitis was well advanced and despite transfer back to the ICU, she died.

It is probable that the leaking anastomosis would have been diagnosed earlier in either the ICU or her surgical ward. The staff of these units, more attuned to such problems, would have noticed the critical symptoms and signs more quickly than happened in the relative backwater of the isolation unit. This might have saved her life.

isolation in each case and who is to define its extent. Attempts to import ready-made solutions are likely to fall foul of local conditions and will almost certainly lead to too much isolation, unnecessary expense and perhaps failure as inadequate resources are overextended. There is no doubt that gut reaction to the subconscious fear of infection keeps alive the mystery and ritual that surrounds much of the practice of isolation. Ethical difficulties make it unlikely that a properly designed clinical trial of isolation itself or even of the methods to be used for it will now take place. The stepwise application of scientifically based common sense seems to be the only answer. At present we seem to be feeling our way towards whatever, in the future, will prove to be the irreducible minimum of isolation practice.

MASKS AND PROTECTIVE CLOTHING

Protective clothing has loomed large in the ritual of infection control. Mythology, the fear of infection, cost consciousness and inadequacy of the scientific data are all interwoven in the decisions that have been made.

Masks

In the medical context face masks are worn to protect patients from microbes expelled from the respiratory tracts of healthcare workers or to protect those who wear them from airborne microbes. In either case the purpose is to prevent infection, though it is difficult for a mask to operate with equal efficiency in both directions. Evidence that they achieve either objective is subjective and anecdotal. In fact, what evidence there is indicates that, when they are not worn, infections among patients do not increase in number. Masks were introduced into surgery at the end of the 19th century, seemingly intuitively rather than for scientific reasons. From operating departments their use has spread to many other areas of healthcare, a tendency that has gone into reverse only recently.

It has been shown that counts of aerial bacteria in ORs are the same whether masks are worn or not (Ritter *et al.*, 1975). A surgical team that operated for a period of six months without masks (432 surgical wounds) noted a reduction in the rate of surgical wound infection compared with the same period in each of the previous four years when masks were worn (Orr, 1981). In an extensive controlled study Tuneval (1991) showed that the wearing of masks made no difference to rates of postoperative sepsis. It has been found that infection rates are unchanged whether or not masks are worn when dressing wounds (Gillespie *et al.*, 1959) and that they are not important in isolation nursing (Ayliffe *et al.*, 1979) or in the delivery room (Turner, Crowley and MacDonald, 1984). In the face of this sort of evidence comparisons between the efficiency of different types of mask as bacterial filters is of no relevance, though some manufacturers use this as a selling point. If they are to be worn for the imagined protection of patients, there seems to be no reason to spend a lot of money on them.

The rationale for this is the fact that very few bacteria are shed from the mouth and nose when breathing and talking normally and even among these few, pathogens are rare (Duguid, 1946). Coughing and particularly sneezing produce somewhat larger numbers, but these are just as likely to emerge round the edges of masks as to be filtered out by passing through them. Because a proportion of the inspired air also enters round the edges of a mask rather than through it (more so as it gets damp) it provides less protection to wearers than they might suppose. It has already been noted that the inevitable rubbing of masks on the face releases potentially infected skin squames (p.147). It is not difficult to imagine where the larger of these end up in the case of a surgeon looking vertically down into a wound.

Rationalization to the point that masks are abandoned altogether is unlikely to be achieved, not least because many of those who are accustomed to wearing them enjoy doing so. However, it should be possible to restrict their use quite drastically and so save money. This would also reduce the need to point out that the improper use of masks is itself a hazard. It would be an advance if staff were no longer seen carrying masks in their pockets, wearing them round their necks or adjusting them with hands that are used immediately afterwards to touch a patient.

The re-emergence of the tubercle bacillus as an iatrogenic hazard for healthcare workers and patients alike has once more thrust face masks into the limelight (p.172). As tuberculosis is one of the few truly airborne infections (p.51) this may have some logic. The intervention of health and safety organizations, whose acknowledged expertise lies in the use of close-fitting masks in industry to protect against the inhalation of particulate matter or chemicals, has added to the confusion that already existed in this biological area (Box 8.3). The requirement for a complete seal between face and mask and the size of the particle concerned (presumed to be a single naked microbe) has suggested solutions not easily applied in clinical situations. The dilemma awaits final resolution, but it does not help that the matter now lies in the field of regulation rather than of scientific enquiry.

Gloves (see also p.178)

As with masks, gloves are used to provide a barrier in two directions, to protect either patients or healthcare workers from pathogenic microbes carried by the other. They may also be used to reduce the unpleasant nature of certain aesthetically unattractive tasks. Gloves are used in different ways to achieve each of these independent objectives.

To protect patients from the microbes on healthcare workers' hands

Box 8.3 Masks, gas masks and industrial respirators

Masks used in war and for various industrial applications are designed to exclude particles, noxious vapours or both. For protection against particles so-called dust-mist or dust-fume-mist respirators are available, but their filters are not made to exclude particles as small as a single naked bacterium. In practice pathogenic bacteria seldom travel singly and are in any case surrounded by residues of the tissues in which they had grown to form a particle much larger than a bacterium on its own. For the protection of healthcare workers against tuberculosis, however, a respirator fitted with a high-efficiency particulate air (HEPA) filter has been specified. This encompasses the possibility that the rod-shaped *Mycobacterium tuberculosis* might travel naked through the air (sideways or end-on?). The large sophisticated respirators needed to achieve this are very expensive and have proved impractical when worn for long periods in clinical situations. A modification was allowed and the specification changed to smaller filters with a lower, 95% efficiency. In mathematical terms this relaxation permits the inhalation of one tubercle bacillus instead of 20 inhaled without a mask! The practical relevance of all this is obscure.

  This is what happens when some event makes it necessary to produce a solution to a problem that science has yet to resolve. A similar process has imposed extreme and extremely expensive clean-room standards (appropriate to the manufacture of miniature electronic components) upon the pharmaceutical industry and upon producers of other sterile equipment: they have also been introduced in some operating rooms.

(their own or transients derived from a previous patient) gloves must be changed or otherwise freed of pathogens at least between patient contacts and sometimes within individual procedures as well. To protect healthcare workers from patients' microbes (provided the gloves are not torn or punctured), they need not be changed or treated in any way at all. To achieve full protection, though, it is important to wash the hands immediately after the gloves have been removed. This is to counter the effect of invisible punctures and because a hand may be contaminated from the outside of a glove as it is removed. All three objectives may be achieved simultaneously if a terminal handwash is added to the changing, washing or antiseptic treatment of gloves as required for the protection of patients. The widespread failure to comply with the requirements for ordinary hand-washing (p.178; Box 8.1) suggests, however, that this additional layer of complexity and possible incomprehension might not improve matters!

  Note that latex and plastic are flammable. It is dangerous to wear gloves close to an open flame. If they ignite they will melt and the burning gloves cannot be removed from the hands.

Protection of the eyes, nose and mouth

It is generally held that mucous membranes are more easily penetrated by microbes than is the case with intact skin. Splashes of body fluids that reach the eyes, nose or mouth are therefore perceived as hazardous, though in the healthcare setting their potential for the transmission of infection is poorly defined. The fear of infection, particularly in its more lethal forms, suggests that this issue is unlikely to be resolved by a controlled trial. Protection is sought by wearing masks, spectacles, goggles or full-face shields. A simple view is that it is sensible to place a barrier between these membranes and the sources of droplets or sprays of body fluids and that the wearing of spectacles (of plain glass for the normally sighted) and an ordinary mask provide adequate protection. Full-face shields are an alternative. Such protection is much more important when power tools are applied to human tissues, as happens in dental surgeries, operating departments and post-mortem rooms. Health and safety regulations may impose different solutions.

Gowns and other protective clothing

A variety of caps, gowns, aprons, boots and other footwear (or disposable coverings for them) have been used in various healthcare settings. Once more they have been employed to provide protection in two directions, for both patients and wearers. Solutions that attempt to satisfy both requirements simultaneously may fail to achieve either.

Fabrics that are comfortable to wear for more than a few minutes are also permeable to quite large particles, so are certainly not microbe-proof. The use of impermeable fabrics or plastic films for short periods when exposed to special risk (a plastic apron, for example) has been a sensible compromise, though if these are single-use items they add cost both at the time of purchase and at their disposal. So far as splashes are concerned (in the context of standard precautions, Table 8.2) a simple clean fabric gown worn for each potential exposure seems a good compromise, though again this has financial implications (see also p.187). This approach is less appropriate in accident and emergency departments, where exposure to blood and body fluids is likely to be more severe and less predictable. The use of special clothing has been proposed for this situation (Steedman, 1994).

Disposable overshoes do not seem to be microbiologically useful in operating departments, rather the reverse (Humphreys et al., 1991a), and they are not thought to be necessary in intensive care units (Gaya, 1980). The headcoverings worn by staff in operating departments do not reduce the numbers of bacteria in the air, so are also of doubtful microbiological value (Humphreys et al., 1991b), though these authors shrink from the suggestion that headcoverings be abandoned for the scrubbed members of surgical teams or at all in orthopaedic operating rooms.

THE RE-USE OF DISPOSABLES

Disposable medical equipment is intended to be used once, then thrown away. In the real world it may be used repeatedly.

Introduction

The title of this section is a contradiction. Disposable medical equipment is made to be used once, then thrown away. Manufacturers and the medical authorities in some richer countries (for different reasons, we hope) are quick to point out that items meant for single use cannot be guaranteed to work properly or indeed may fail structurally if used more than once and that it may be difficult or impossible to clean and if necessary resterilize them between uses. If the apparatus has a volumetric function, they say that accuracy may be lost. Finally they highlight the fact that if an accident occurs as the result of re-using a disposable the fault lies with the user, who must bear the consequences, including the cost of any litigation.

As medicine has become more technical, the range and complexity of the equipment necessary to it has extended enormously. At one time much of it was made to be re-used. A revolution followed the appearance of improved plastics and the skills needed to manipulate them. Commercial and other pressures led to cost-effectiveness studies that showed that money could be saved by replacing some re-usable equipment with single-use disposable items made of plastic. These studies may have left out certain costs, such as that of providing the extended facilities required to transport and store the vastly increased volume of disposables or the cost of getting rid of them after use. They certainly took no account of the damage their disposal does to incinerators or to the environment (p.203). Today much new equipment is disposable from the outset. Most of it is attractive to look at and pleasant to handle, but the cost of it is now an important part of healthcare budgets. Even in the richest countries some of the more expensive 'single-use' items such as cardiac catheters, pacemakers and haemodialysis coils are recycled. In the poorest countries each 'disposable' syringe may be used up to 40 times before it is discarded. In the middle a wide range of equipment is reprocessed, including items such as eye shields, oxygen masks, airways, anaesthetic tubing, tracheostomy and endotracheal tubes and foetal scalp electrodes. There is a wide gap between what the manufacturers would like and what really happens.

Recycling single-use items

If a hospital decides to re-use single-use items, the decision should be made consciously in each case, with full consultation that should include the infection control organization. Each step of the procedure necessary to recycling, including cleaning, examination for structural integrity and, if appropriate, testing for function, the type of any packaging required, the method used for sterilization or disinfection and how to keep track of the number of times each item has been reprocessed, should be defined and written down. The work should be supervised by a responsible person and it may be integrated into hospital sterilizing and disinfecting units (p.240).

As with any disinfection or sterilization process, the preliminary wash is of vital importance. This is particularly true of tubes, more so if they are long and thin and might contain debris or blood that has clotted. It may be necessary to devise a special flushing system in these cases and to show that the tube is fully patent before it is processed any further. The final flush, at least for items used parenterally, should be with apyrogenic distilled water or an equivalent. Washing is followed by drying, perhaps with warm air, but in any case in an atmosphere free at least of larger particles of dust.

The options for sterilization or disinfection include autoclaving, ethylene oxide, perhaps free radicals, low temperature steam, hot water pasteurization and exposure to liquid disinfectants, of which glutaraldehyde, formaldehyde, alcohol and hypochlorite have been used more commonly (p.120). Among the plastics polystyrene will barely stand up to pasteurization, though with care PVC (vinyl) will. Polypropylene, PTFE and epoxy and silicone resins will all withstand autoclaving. For less critical items that will only come into contact with non-sterile mucous membranes, pasteurization is an adequate process. Liquid chemicals should only be used when there is no alternative.

Accidents resulting from the re-use of single-use equipment are not in evidence, though they must have occurred: the lack of information is not surprising. Provided recycling is performed responsibly, however, there is no reason why it should add perceptibly to the risks already inherent in medical treatment. Much of the population of the world has no access to modern medicine and even in the richest countries politicians can no longer conceal the existence of rationing. If the money saved by recycling disposables is used to help people who would otherwise be deprived of medical care, then so be it.

WASTE AND ITS DISPOSAL

The disposal of clinical waste has become a public issue. Its generation, segregation and disposal are discussed.

Introduction

It is a common human perception that waste is potentially and often actually offensive and that it is a health hazard. This perception is strengthened when the waste arises in a healthcare setting. Media reports of the finding of such things as used syringes or even human tissues in refuse tips or washed up on beaches has turned this into a major public issue. Hospital administrators must pay proper attention to the disposal of their waste or they risk adverse publicity, humiliation and perhaps legal proceedings. As with other aspects of the management of healthcare, the disposal of clinical waste has become more a matter of expensive regulation than of common sense.

As often happens, public perception has linked the unaesthetic with the hazardous. With certain specific exceptions, the mass of waste generated in medical establishments is certainly no more dangerous than that from ordinary households. In hospitals wet food waste (with its potential for supporting bacterial multiplication) is usually separated from the inherently much safer dry waste, so what emerges for disposal is often less objectionable in a microbiological sense than household waste. Even if this were not so the distinction between reservoirs and sources of infection needs to be made (p.43). With the vital exception of penetrating injuries due to bloodstained sharp objects and exposure to excrement, instances of infection arising from exposure to rubbish, either from hospitals or the rest of the community, are not evident. Short of someone physically rolling about in rubbish, it is difficult to see how any microbes it may contain could cause an infection.

Categories of waste

Waste from medical establishments ought to be categorized into different classes if the most stringent (and therefore most expensive) disposal procedures are not to be applied to the whole of it. The particularly obnoxious and perhaps dangerous byproducts of the healthcare industry ('clinical waste') for which special disposal methods are necessary may not exceed 10% of the total. The proportion might be reduced still further if more precise definition and discipline were applied to what constitutes truly dangerous waste. A working classification of waste is provided in Table 8.3 and methods appropriate to the disposal of each category are noted. For completeness excremental waste (faeces and urine) is included.

Excremental waste

In places where a public water-carried sewage system is not available, a considerable problem may arise with the disposal of faeces and urine. If disposal can be arranged by water carriage into properly designed and maintained private sewage systems or septic tanks, then the situation is retrieved. This is not always possible, however, and unfortunately it is precisely in these places that diseases with a faecal–oral route of transmission may be much more common. Here patients with enteric fever (typhoid or paratyphoid), dysentery, cholera or some forms of hepatitis are often found in hospitals. Unless one of the hygienic means for the disposal of faeces and urine described above can be provided, such patients are a hazard to other patients, to staff and to the community in the neighbourhood. The addition of a disinfectant to bedpans is not a complete answer, for two reasons.

Table 8.3 Categories of waste that may arise in medical establishments and how they may be disposed of

Categories of waste	Disposal method
Excremental and other liquid waste	Water-carried sewage systems, when available (see text)
Domestic rubbish (trash)	Same method as used in the surrounding community for household and food waste, with recycling and composting of as much of it as possible
Clinical waste • offensive or microbiologically hazardous waste* • human tissues • microbiologically hazardous laboratory waste, after autoclaving in-house	Incineration or such other methods as are approved by national regulatory or other responsible bodies
• all disposable sharp objects	Direct disposal into special sharps container (see text). When full, these are incinerated or dealt with as above
• Residues of certain particularly toxic drugs or radioactive materials	As required by national regulatory bodies
Other waste (confidential documents, pharmaceuticals, flammable materials, etc.)	May be incinerated or as determined locally

* Used swabs, dressings, disposable clothing, incontinence pads, disposable suction bottles, sputum pots, etc.

First, the amount of disinfectant is unlikely to be sufficient to do what is necessary and second, the toxic phenolic disinfectants commonly used contaminate the environment and may find their way back into drinking water.

The disposal of 'sharps'

Special containers for the disposal of sharp objects ('sharps containers') are necessary because of the unacceptably high incidence of accidents to hospital employees. In many cases these have suffered needle-stick and other injuries from the growing number of disposable sharp objects when they handle bags of 'ordinary' waste into which, improperly, they have found their way. Infection with hepatitis B has followed such accidents and the possibility of HIV infection is a newer, though fortunately less serious hazard. Sharps containers need to be rigid and waterproof and made of material that cannot easily be penetrated by, for example, a hypodermic needle. Ideally the top opening should be arranged so that once inserted, an object cannot be removed or escape if the container is upset. When full, it should be possible to close this opening in a positive and irreversible fashion, for safe disposal (Gwyther, 1990).

Disposable syringes and needles have been responsible for most of the accidents in which healthcare workers have been pierced by potentially bloodstained sharp objects. A contentious issue has been the practice of capping needles with their original sheaths after they have been used. If recapping is banned the primary user may be protected from an accident, but others involved in the later stages of the disposal process are at risk of exposure to unguarded, contaminated sharp points. Recapping can be done safely by a technique that keeps hands (including those of bystanders) behind the needle at all times or by the use of specially designed syringes in which needles are covered after use by sleeves that slide forward from the barrel of each syringe.

Additional to any such measures, sharps containers are used. Opinions differ on whether a syringe should be discarded with the needle attached or if the needle should be put into the container on its own and the syringe discarded elsewhere. Some containers are equipped with keyholes into which the hub of a needle can be fitted, to allow it to be twisted off so the needle falls inside without risk to the hands of the operator. Such keyholes tend to become bloodstained. The argument may turn on the cost of sharps containers and the speed with which they are filled. In places where the cost of special sharps containers is prohibitive, an alternative is to use the plastic containers in which liquids (disinfectants, for example) have been supplied. In any case containers should not be filled to the top. If something is forced into an already overfilled container, somewhere a needle may be pushed through the side of it. Sharps containers do not need to be discarded on a regular basis, but only when they are comfortably full.

Incinerators

For a long time incineration has been seen as the ideal way to dispose of at least the clinical category of waste and sometimes all of it (p.115). This is no longer universally true and the growth in the amount of disposable plastic used in hospitals is one of the chief problems. This has significantly increased the volume of waste to be dealt with, while it only marginally increases its weight. At one time, when an incinerator was installed its size was determined by the weight of material to be burnt rather than its volume, so many of them are now too small. This is not all. When they are burnt some plastics produce dense black smoke plus significant amounts of hydrochloric acid and smaller quantities of other more toxic compounds. All of them produce carbon dioxide. The smoke is unacceptable to communities with clean air acts, the carbon dioxide contributes to the 'greenhouse effect', the acid eats away the inside of the incinerator before it becomes acid rain and the toxins are environmentally very unfriendly. Some of these problems may be overcome by using more sophisticated and expensive incinerators, but international concern for the environment has already led some countries severely to limit their use. This attitude is likely to spread. Alternative methods for rendering clinical waste both safe and aesthetically acceptable have been designed, but they are expensive, depend on complex machinery that will break down from time to time, use toxic chemicals and produce a bulky endproduct that still has to be disposed of.

Landfill

The usual alternative to incineration is landfill. In overpopulated or environmentally very conscious countries it has become increasingly difficult to find sites where this is socially or ecologically acceptable. Perhaps more seriously, there is growing concern about the leakage of toxic materials from such dumps. In some cases these have already reached underground sources of drinking water and in others it is feared that this will happen in the future. Unless special collection systems are installed to carry them away safely, methane and other gases escape into the air as some kinds of rubbish undergo microbial degradation.

Hospital waste

The problems that afflict hospitals when they try to dispose of their waste are no more than rather acute reflections of a difficulty that increasingly involves the whole human race. The 'effluent society' that inhabits the more affluent parts of the globe has become a cliché, but it cannot be long before we are forced to do something about it. Pressure to do so comes from three sides. First is the already difficult problem of what to do with today's waste. Second, as the population of the world expands the 'effluent' group gets larger, its waste more voluminous and the problem will grow. Third is the finite nature of the earth's resources. In the case of disposable plastics, hydrocarbons that have taken millions of years to form are converted into an item that is used once and then burnt to convert it back into carbon dioxide or it is put into a hole in the ground where it may remain intact for years. In due course exhaustion of the earth's store of hydrocarbons or global warming, or both, will put an end to this. As a first step manufacturers ought to be persuaded to stop wrapping items in packages that are often many times larger than they need be and may weigh more than the item itself. They should also recycle their packaging and other plastic materials and purchasers should encourage this. Medical services might eventually be driven to reverse the move towards making everything disposable. For example, it would not be impossible to revert to the use of reusable glass syringes.

The infection control committee or whatever body is appointed for the purpose in each hospital ought to take these factors into account as they produce new or modify existing waste disposal policies. At the same time supplies organizations should develop an environmental conscience and press manufacturers to remove the rubbish they generate. So far as waste is concerned, it is necessary to decide on the number of categories to be identified and arrange for the supply of suitable bags or other containers in an equivalent number of distinctive colours to pack it in. Then, taking account of local conditions and probably of existing regulations, the method of disposal is chosen. Incineration at the hospital or at a distant shared site are options. When a distant site or landfill are used secure means of transport and storage are required. These must be vermin- and 'little boy'-proof.

One of the most important parts of the whole system is a training programme designed so that all members of staff know and understand the hospital policy and are motivated to adhere to it. Finally, the system ought to be monitored actively (audited) to correct errors before these become matters of public concern.

REFERENCES

Albert, R. K. and Condie, F. C. (1981) Hand-washing patterns in medical intensive-care units. *New England Journal of Medicine*, **304**, 1465–6.

Ayliffe, G. A. J. (1992) Efficacy of handwashing and skin disinfection. *Current Opinion in Infectious Diseases*, **5**, 542–6.

Ayliffe, G. A. J., Babb, J. R., Taylor, L. *et al.* (1979) A unit for source and protective isolation in a general hospital. *British Medical Journal*, **ii**, 461–5.

Centers for Disease Control (1988) Update: universal precautions for prevention of transmission of human immunodeficiency virus, hepatitis B virus, and other bloodborne pathogens in health care settings. *Morbidity Mortality Weekly Report*, **37**, 377–88.

Duguid, J. P. (1946) The size and duration of air carriage of respiratory droplets and droplet nuclei. *Journal of Hygiene, Cambridge*, **44**, 471–9.

Garner, J. S. (1996) Guidelines for isolation precautions in hospitals. *Infection Control and Hospital Epidemiology*, **17**, 53–80.

Garner, J. S. and Simmons, B. P. (1983) CDC guidelines for isolation precautions in hospitals. *Infection Control*, **4**, 245–325.

Garner, J. S. and Simmons, B. P. (1986) Isolation precautions, in *Hospital Infections*, 2nd edn, (eds J. V. Bennett and P. S. Brachman), Little, Brown, Boston, pp. 143–50.

Gaya, H. (1980) Is it necessary for staff and visitors in intensive care units to wear masks, hats, gowns and overshoes? *Journal of Hospital Infection*, **1**, 369–71.

Gillespie, W. A., Alder, V. G., Ayliffe, G. A. J. *et al.* (1959) Staphylococcal cross-infection in surgery. Effects of some preventive measures. *Lancet*, **ii**, 781–4.

Gwyther, J. (1990) Sharps disposal containers and their use. *Journal of Hospital Infection*, **15**, 287–94.

Haley, R. W., Tenney, J. H., Lindsey, J. D. *et al.* (1985) How frequent are outbreaks of nosocomial infection in community hospitals? *Infection Control*, **6**, 233–6.

Holt, H. M., Anderson, S. S., Gahrn-Hansen, B. *et al.* (1995) Infections following epidural catheterisation. *Journal of Hospital Infection*, **30**, 253–60.

Howells, C. H. L. and Young, H. B. (1966) A study of completely undressed surgical wounds. *British Journal of Surgery*, **53**, 436–9.

Humphreys, H., Marshall, R. J., Ricketts, V. E. *et al.* (1991a) Theatre overshoes do not reduce operating theatre floor bacterial counts. *Journal of Hospital Infection*, **17**, 117–23.

Humphreys, H., Russell, A. J., Marshall, R. J. *et al.* (1991b) The effect of surgical head-gear on air bacterial counts. *Journal of Hospital Infection*, **19**, 175–80.

Jansen, B. and Peters, G. (1991) Modern strategies in the prevention of polymer-associated infections. *Journal of Hospital Infection*, **19**, 83–8.

Jarvis, J. D., Wynne, C. D., Enwright, L. *et al.* (1979) Handwashing and antiseptic-containing soaps in hospital. *Journal of Clinical Pathology*, **32**, 732–7.

Kurtz, J. B. and Boxall, J. (1976) A partial substitute for hand washing. *Nursing Times*, **72**, 332–3.

Larson, E. and Killian, M. (1982) Factors influencing handwashing behavior of patient care personnel. *American Journal of Infection Control*, **10**, 93–9.

Lynch, P., Jackson, M. M., Cummings, M. J. *et al.* (1987) Rethinking the role of isolation practices in the prevention of nosocomial infections. *Annals of Internal Medicine*, **107**, 243–6.

Maki, D. G. (1986) Skin as a source of nosocomial infection. Directions for future research. *Infection Control*, **7** (supplement), 113–16.

Mathews, J. A. and Newsom, S. W. B. (1987) Hot air electric hand driers compared with paper towels for potential spread of airborne bacteria. *Journal of Hospital Infection*, **9**, 85–8.

Meers, P. D. and Leong, K. Y. (1989) Hot-air hand driers. *Journal of Hospital Infection*, **14**, 169–71.

Meers, P. D. and Leong, K. Y. (1990) The impact of methicillin- and aminoglycoside-resistant *Staphylococcus aureus* on the pattern of hospital-acquired infections in an acute hospital. *Journal of Hospital Infection*, **16**, 231–9.

Ojajarvi, J., Makela, P. and Rantasalo, I. (1977) Failure of hand disinfection with frequent hand-washing: a need for prolonged field studies. *Journal of Hygiene, Cambridge*, **79**, 107–19.

Orr, N. W. M. (1981) Is a mask necessary in the operating theatre? *Annals of the Royal College of Surgeons of England*, **63**, 390–1.

Preston, G. A., Larson, E. and Stamm, W. E. (1981) The effect of private isolation rooms on patient care practices, colonization and infection in an intensive care unit. *American Journal of Medicine*, **70**, 641–5.

Reybrouck, G. (1986) Handwashing and hand disinfection. *Journal of Hospital Infection*, **8**, 5–23.

Ritter, M. A., Eitzen, H., French, M. L. V. *et al.* (1975) The operating room environment as affected by people and the surgical face mask. *Clinical Orthopedics and Related Research*, **111**, 147–50.

Sanderson, P. J. and Weissler, S. (1992) Recovery of coliforms from the hands of nurses and patients: activities leading to contamination. *Journal of Hospital Infection*, **21**, 85–93.

Steedman, D. J. (1994) Protective clothing for accident and emergency personnel. *Journal of Accident and Emergency Medicine*, **11**, 17–19.

Tuneval, T. G. (1991) Postoperative wound infections and surgical face masks: a controlled study. *World Journal of Surgery*, **15**, 383–8.

Turner, M. J., Crowley, P. and MacDonald, D. (1984) The unmasking of delivery room routine. *Journal of Obstetrics and Gynaecology*, **4**, 188–90.

Williams, E. and Buckles, A. (1988) A lack of motivation. *Nursing Times*, **84**, 60–4.

Zimakoff, J., Stormark, M. and Olsen Larsen, S. (1993) Use of gloves and hand-washing behaviour among healthcare workers in intensive care units. A multicentre investigation in four hospitals in Denmark and Norway. *Journal of Hospital Infection*, **24**, 63–7.

FURTHER READING

Ayliffe, G. A. J. (1988) Equipment-related infection risks. *Journal of Hospital Infection*, **11** (supplement), 279–84.

Babb, J. R. (1988), Methods of reprocessing complex medical equipment. *Journal of Hospital Infection*, **11** (supplement), 285–91.

Kolmos, H. J. (1983) A clinical and bacteriological evaluation of a re-use system for disposable haemodialysers. *Journal of Hospital Infection,* **4**, 269–78.

Sproat, L. J. and Inglis, T. J. J. (1994) A multicentre survey of hand hygiene practice in intensive care units. *Journal of Hospital Infection*, **26**, 137–48.

9 Infection control in the community

IATROGENIC INFECTIONS IN THE COMMUNITY

Ageing populations and growth in the volume of healthcare delivered outside hospitals has increased the number and importance of the iatrogenic infections found in the community. Their causes and control are discussed.

Introduction

Hospitals are integral parts of the communities in which they are sited. Not only do they provide specialist medical care for those who need it, but they are also major employers. Large numbers of individuals, staff and patients, move daily across the complex interface between every hospital and the community it serves. The interface must accommodate activities that range from community support (fund-raising activities, leagues of friends, voluntary services) to areas of potential or real friction (transport, disposal of hospital waste, atmospheric pollution and the transmission of infection by staff or patients). For the reasons discussed in Chapter 1 problems associated with the transfer of infections between hospitals and the communities they serve are of growing importance.

Changing patterns in the provision of healthcare have blurred the former clearcut distinction between infections that were acquired in the community (community-acquired infections, CAI) and those that appeared as a result of admission to hospital (hospital-acquired infections, HAI). Economic imperatives are shifting the centre of gravity of healthcare into the community (Chapter 1). We have acknowledged this change of emphasis by the use of the term 'iatrogenic infection' (IaI) to expand the meaning of HAI. The counterpart of IaI is community infection (CI). CIs complement IaIs so that, as was the case with HAI and CAI, the total of IaI plus CI equals the whole of infectious disease.

IaIs make only a small contribution to the whole of infectious disease. The more common CIs range from the most prevalent variety, dental caries, through (as examples) boils, styes, whitlows, septic abrasions, coughs, colds, 'flu and other respiratory infections, urinary tract infections, outbreaks of food poisoning, foodborne hepatitis and dysentery, to the childhood fevers, bloodborne hepatitis and HIV infection, plus the infestations. The diagnosis, treatment and control of these CIs are among the bread-and-butter day-to-day activities of doctors and other healthcare workers in the community.

IaIs, by definition, result from the delivery of healthcare, so strictly speaking are only possible in circumstances in which healthcare is delivered. This again blurs the boundary that separates IaIs from CIs. At what point, for example, does a home for the elderly become a healthcare facility? Many of them offer sheltered accommodation to elderly people with no more than minor disabilities. New residents may require no special healthcare provision so at this stage and strictly speaking, any infections they suffer are CIs. Later when more medical support is needed their infections may become IaIs. To supply the need, homes are visited on a regular basis by doctors and community nurses or they may have trained nurses or nursing auxili-aries on their staffs. In these circumstances the definition of IaI may be simplified by accepting that, as nursing homes extend hospital care into the community, any infections among their residents are counted as IaIs except as otherwise determined. Those who work in the field may wish to use other definitions, but general agreement is important if meaningful comparisons are to be made. An acceptable definition must encompass, for example, the case of an outbreak of influenza in a nursing home. Are cases of it to be counted as CIs or, if the medical advisor has failed to recommend vaccination, as IaIs or what?

Some nursing homes provide more elaborate services for the severely disabled on a permanent or respite basis. Special units that provide day or residential care for children, the younger chronic sick, those with AIDS or psychogeriatric problems or the terminally ill further confuse the picture. These major variables make it difficult to identify reliable data about the extent of IaI in the community. Differ-ences in the types of resident accommodated and the ranges of services offered mean that the results of surveys in different types of facility cannot be compared (see, for example, Danowski, Gordon and Simor, 1991).

Most of the cases of IaI that arise in hospitals are the result of various kinds of sustained invasive procedure. With the exception of urinary catheterization and a small but growing number of other types of long-term intubation, these are less commonly performed in the community. Despite their relative infrequency, however, the IaIs that complicate these and other procedures must not be ignored. They are important precisely because they *are* iatrogenic. As unwanted, accidental and to an arguable extent avoidable byproducts of the delivery of healthcare, IaIs must engage the attention of those who have caused them or, at the least, have failed to prevent them. All healthcare facilities need to have a written set of procedures for the control of infection, either as a section of a larger policy manual or as a separate document. In the modern world failure to undertake this task voluntarily risks an enforced change of attitude at the hands of regulators or of lawyers and their clients.

So far as IaIs are concerned they are recognized and prevented in the same way, whether they arise in the community or in hospitals. It is only the circumstances that differ, so the principles and practices described in the rest of this book are relevant in either setting. To avoid repetition, reliance is placed on cross-references. We hope those whose interests are centred in the community will excuse this.

This book is concerned with IaIs. Community healthcare workers who need to be better informed about CIs must refer to more general works on infection (see Further Reading on p.214). The administrative arrangements made for the care of these infections vary from place to place. In the UK responsibility lies with consultants in communicable disease control (CCDCs) and the officers of Environmental Health Departments. Members of both groups also have some responsibilities within hospitals. Whatever the arrangement, however, infection control organizations need to be fully integrated across what, without proper consultation and co-operation, has a potential to develop into an administratively awkward, unhealthy gap between the services offered in hospitals and in the community.

Nursing homes (long-term care facilities)

Changes in the way healthcare is delivered are particularly well seen in the practice of geriatric medicine. Ageing populations, a reduction in the number of the expensive hospital beds set aside for long-term care of the elderly, a growth in the number of private nursing homes and an increased availability of community support in individuals' own homes have each had an effect. New residents who enter nursing homes today are significantly older and more frail than was formerly the case (Gosney, Tallis and Edmond, 1990/91). The result is a requirement for more skilled nursing and medical attention in the community.

Surveys of infection in some types of nursing home have revealed prevalences of infection of about 15% (Garibaldi, Brodine and Matsumiya, 1981; Bentley and Cheney, 1990; Garibaldi and Nurse, 1992). The types of infections found vary and they differ from the distribution found in hospitals. A not uncharacteristic pattern would be infected pressure sores (6%), conjunctivitis (3.4%), respiratory infections (2.8%), urinary tract infections (UTIs, 2.6%) and gastro-enteritis (1.3%). Colonizations and infections rise to 100% among patients catheterized for extended periods, so the prevalence of UTI varies directly with the frequency of long-term catheterization in the home surveyed and this in turn varies with the type of person cared for. The microbial pathogens involved reflect the chronic nature of these infections (p.130). Latterly MRSA, *Clostridium difficile* and, in some settings, *Mycobacterium tuberculosis* have emerged as problems. In some circumstances hepatitis and HIV infection may be included.

Insofar as they are preventable, the control of these infections is by the methods described elsewhere in this book. A special issue in infection control in the community is associated with the 'problem' organisms named at the end of the last paragraph. Many nursing home residents need to be admitted to hospital from time to time. While there, they may be colonized or infected by undesirable organisms, of which MRSA has attracted a good deal of attention (Working Party, 1995). Some nursing homes find themselves ostracized by hospitals or other nursing homes when it is known that one or more of their residents are carriers of MRSA. Such carriage can cause difficulties with the insurances carried by private nursing homes. To avoid these difficulties some of them refuse to accept former or potential new residents if they are or even might be carriers of MRSA. Expensive acute hospital beds can be blocked when such patients cannot be discharged although they are fit for it. Issues of these kinds can usually be resolved by discussion between the parties concerned, with advice from local infection control organizations. The discussions tend to be time consuming, however, and the incidents themselves are disruptive and distressing to patients and their relatives. They contribute significantly to the demonization of 'superbugs' (see Box 11.1, p.235)

Outbreaks of respiratory infection and gastroenteritis are features of life in nursing homes. Both are significant causes of morbidity and mortality among frail elderly residents. They are commonly viral in origin, but gastroenteritis may also be bacterial (p.161). Among the respiratory viruses influenza viruses A and B are prominent as pathogens and prophylactic vaccination is advised.

Child day care centres

Changing patterns of employment are responsible for a great increase in the need for child day care services. Outbreaks of infection are inevitable when groups of similar individuals are brought together for significant periods and children in day care centres are no exception. Pre-school infants and children are at even greater risk in this respect as many of them have yet to develop their own natural immunity to the whole range of childhood infectious diseases. Gastroenteritis, otitis media, varicella, hepatitis A and the childhood fevers are all common (Chouillet, Maguire and Kurtz, 1992; Thacker et al., 1992). Requirement for vaccination and policies for the temporary exclusion of children who have been exposed to infections are both important issues. Although the management of infections and infestations in these centres may involve infection control organizations, only the most liberal of definitions would classify individual cases of such infections as iatrogenic. The decision is more difficult in an outbreak and once more, generally agreed definitions are needed.

Other healthcare facilities in the community

A major proportion of healthcare delivery takes place in general practitioners' health centres and in dental surgeries. A patient rarely spends more than a few minutes with a doctor, dental surgeon or nurse and perhaps a little longer in a waiting area. To some extent individuals with infections congregate in doctors' and dentists' waiting rooms. The transmission of infections between patients is a possibility, but in this respect they differ little from other communal meeting places such as churches, shops and places of recreation or amusement. It is what happens in the consulting room, treatment room or dental chair that determines if the delivery of healthcare is accompanied by the transmission of IaIs from patient to patient or between patients and healthcare workers, in either direction.

It has been suggested that, in general, fewer invasive procedures are performed in community as compared with hospital medical practice. This is clearly not true of dental surgery, however and in medical health centres endoscopic examinations and minor surgery are now more frequent occurrences. In these circumstances the principal hazards are the transmission of pathogenic microbes from one patient to the next on the surfaces of inadequately sterilized or disinfected equipment (p.106) or to a healthcare worker who is exposed to blood or body fluids of an infected patient. These risks are particularly acute in dental surgeries. Hypodermic needles and sharp instruments are in constant use. In addition high-speed dental machinery produces large numbers of small particles that are projected over short distances at considerable speeds, possibly to strike the hands and faces of dental surgeons and chairside assistants. Excellent specialist advice is available on the management of these risks (see, for example, Wood, 1992), but it seems that the sensible precautions proposed are not always adopted (Scully et al., 1994).

In hospitals reusable equipment is cleaned and disinfected or sterilized according to fixed protocols by identified staff who may be specialists in this activity. Their work is subject to scrutiny and often to formal quality control. Mistakes still happen, but they are rare. By comparison disinfection and sterilization as practised in small health-care facilities in the community are less controlled, so more open to error. A single nurse or a nurse receptionist has many tasks to perform, only one of which may be to care for whatever equipment is reused. Staff changes may be frequent and these seemingly mundane tasks may be afforded low priority, the facilities for cleaning instruments prior to disinfection or sterilization may be inadequate and education is easily neglected. The processes are more likely to be poorly understood and badly performed. Faithful adherence to written instructions supplied by the manufacturer of the equipment and to advice from professional organizations is a great help but they are not a substitute for advice and, better, visits from experts from local infection control organizations. Contacts should be made on a regular basis.

Attention to infected wounds (p.184) is a regular and growing feature of community nursing. The necessary care is often delivered in individual patients' homes so there is (or should be) little opportunity for cross-infection. This does not apply when dressings are changed in health centre treatment rooms. Nurses who work in health centres and in the community easily become isolated and in these circumstances standards may, in time, be eroded. The remedy once more is continuing education, supplemented by advice from local infection control experts.

Occupational health in community healthcare

Healthcare workers in the community see large numbers of patients of whom little or nothing may be known until the time of the consultation. For many, occupational exposure to infection is a daily occurrence. In most cases and for most infections the risk is accepted and healthcare workers probably do not suffer from the common minor infections more often than employees of other organizations whose jobs involve regular direct contact with members of the public. However, this does not extend to certain serious or life-threatening infections. Hepatitis B and C, infection with the human immunodeficiency virus (HIV) and tuberculosis are the most important of these (pp.164 and 233).

Before hepatitis B vaccine was available hepatitis B infection was a serious occupational hazard for healthcare workers. It is important for all who come into close physical contact with patients to be vaccinated. There is as yet no HIV vaccine, but fortunately this infection is much less easily transmitted than hepatitis B, so occupational HIV infections have been very rare and most are avoidable (p.236). The protection of the hands from sharp bloodstained objects, the wearing of gloves and, if sprays or splashes are likely, the use of spectacles, goggles, masks or face shields are recommended (p.195). As noted elsewhere, tuberculosis is another matter. Advice on this subject is still under development but, as usual, overreaction to the problem has been the initial response (p.172). The use of BCG vaccine must be considered.

REFERENCES

Bentley, D. W. and Cheney, L. (1990) Infection control in the nursing home: the physician's role. *Geriatrics*, **45**, 59–66.

Chouillet, A., Maguire, H. and Kurtz, Z. (1992) Policies for control of communicable disease in day care centres. *Archives of Diseases in Childhood*, **67**, 1103–6.

Danowski, S. B., Gordon, M. and Simor, A. E. (1991) Two years of infection surveillance in a geriatric long-term care facility. *American Journal of Infection Control*, **19**, 185–90.

Garibaldi, R. A. and Nurse, B. A. (1992) Infections in nursing homes, in *Hospital Infections*, 3rd edn, (eds J. V. Bennett and P. S. Brachman), Little, Brown, Boston, pp. 491–505.

Garibaldi, R. A., Brodine, S. and Matsumiya, S. (1981) Infections among patients in nursing homes. Policies, prevalence, and problems. *New England Journal of Medicine*, **305**, 731–5.

Gosney, M., Tallis, R. and Edmond, E. (1990/91) The burden of chronic illness in local authority residential homes for the elderly. *Health Trends*, **22**, 152–7.

Scully, C., Blake, C., Griffiths M. *et. al.* (1994) Protective wear and instrument sterilisation/disinfection in UK general dental practice. *Health Trends*, **26**, 21–2.

Thacker, S. B., Addiss, D. G., Goodman, R. A. *et al.* (1992) Infectious diseases and injuries in child day care. *Journal of the American Medical Association*, **268**, 1720–6.

Wood, P. R. (1992) *Cross Infection Control in Dentistry*, Wolfe, London.

Working Party (1995) Guidelines on the control of methicillin-resistant *Staphylococcus aureus* in the community. *Journal of Hospital Infection*, **31**, 1–12.

FURTHER READING

Mandell, G. L., Bennett, J. E. and Dolin, R. (1995) *Principles and Practice of Infectious Diseases*, 4th edn, Churchill Livingstone, New York.

Meers, P., Sedgwick, J. and Worsley, M. (1995) *The Microbiology and Epidemiology of Infection for Health Science Students*, Chapman & Hall, London.

Infection control in other special environments | 10

OPERATING DEPARTMENTS

Operating departments are expensive. Some costs represent an overelaboration of the measures used to control infection. Money can be saved if unnecessary precautions are modified or abandoned.

Infection, sepsis, antisepsis and asepsis

Many of the things done to prevent surgical sepsis were introduced intuitively or by extension from inadequate or misinterpreted data (see pp.135, 184 and 195). Operating rooms (ORs) are places where something called asepsis is practised, amid much ritual.

The two major English language dictionaries define sepsis as a toxic condition resulting from the multiplication of pathogenic bacteria and their products in the region of an infection or as a state of poisoning of the tissues or bloodstream caused by bacteria. An infection is the result of an interaction between two living things, a host and a parasite. It follows that something inanimate or abstract cannot be infected, septic or 'aseptic', yet even medical dictionaries use the words as if this were possible. When Lister invented antiseptic surgery he imagined an attack on what he thought were the airborne causes of sepsis, so in this context the term 'anti-sepsis' was correctly applied. When surgery evolved to become 'aseptic' the word acquired a new meaning. Aseptic surgery could accurately describe an ideal form of surgery free from infection. In fact, the word is often used to describe surgery or other activities performed in the (always imaginary) absence of microbes.

Though well established and probably irreversible, it is necessary to point to the confusion this usage has caused. True asepsis is as difficult to define as sterility (p.107) with the added need to decide whether something is aseptic if it is contaminated with dead, though perhaps still toxic bacteria. To highlight the problem, imagine a nurse shedding some 2000 bacteria-carrying skin squames every minute (p.45) wiping away or covering up a few hundred bacteria on a surface to prepare an 'aseptic field' to perform an 'aseptic procedure' on a patient with an infected (septic) wound. Fields and procedures cannot be either infected or septic, so how can they be aseptic?

Meanwhile both nurse and patient are home to a normal flora of some 10^{14} (1 and 14 zeros) individual bacteria. Even when the word is used incorrectly, the only truly aseptic technique would be one carried out inside an enclosure completely free of microbes in which one sterile robot operated on another. It is not surprising that such a conveniently imprecise term has become a vehicle for woolly thinking and has been surrounded by rhetoric.

Microbes in operating departments

With modern technology it is (expensively) possible to achieve near-sterility in all parts of the environment surrounding the tissues of a patient exposed during surgery. It is inconvenient that neither these tissues nor the bodies of the patient and staff present in the OR can be sterilized. Virtually all the bacteria present in the air or on surfaces in ORs are derived from the bodies, chiefly the skin, of its occupants. Bacteria are shed continuously from skin (p.45), aided by movement and rubbing of clothes and at an increased rate for some time after bathing or washing the hands (Speers *et al.*, 1965; Meers and Yeo, 1978). Normal theatre garb is no more and may even be less of a barrier to the escape of these bacteria than normal clothing (Bethune *et al.*, 1965; May and Pomeroy, 1973; Dankert, Zijlstra and Lubberding, 1979; Laufman *et al.*, 1980; p.195). After a short time theatre dress is as prolific a source of microbes as were the outdoor clothes recently removed or possibly more so if the person has just taken a shower. There is no microbiological reason for requiring theatre staff to change their clothes or for wearing what has become conventional OR garb.

'Clean' and 'dirty'

For the same reason there is no microbiological reason to distinguish between members of staff who perform 'clean' and 'dirty' tasks. Everyone carries a similar load of microbes and unless they are enclosed in an uncomfortable all-enveloping suit made of impervious material and wear a space-style helmet, all share the habit of shedding them into their surroundings. In numerical terms a dirty object handled with reasonable care is most unlikely to create more environmental contamination than that generated at the same time by the skin of the person performing the task. Of course, some individuals are more active disseminators of bacteria and this becomes important if the source is an infected condition, because the significantly larger number of bacteria released are also more pathogenic (pp.42 and 43).

Microbes are effectively immobile, so depend on other agencies to carry them about. In ORs, as in any occupied enclosed space, the most important of these agencies are human beings and direct contact between them the best way of effecting their transfer. Transfer through the environment is very inefficient and for most microbes the environment is hostile. Dry dust contains few living bacteria and the longer it has lain, the fewer there are. Really heavily contaminated inanimate environmental items (a suction bottle containing pus, for example) are rare. Provided they are securely enclosed in sealed containers or placed inside an impervious wrapper and are correctly disposed of, even heavily contaminated items pose no threat. There is no microbiological reason why contaminated and non-contaminated items or so-called 'clean' and 'dirty' people should not pass each other in the same corridor. The provision of 'clean' and 'dirty' corridors in operating departments has no microbiological justification.

Transfer zones for the passage of patients, goods and staff into and out of operating departments are required for functional rather than microbiological reasons. People are the main source of bacteria, so the number of people admitted to the department (and their activity) should be kept to a minimum. This requires discipline. Transfer zones help to impose this. It is incorrect to think the outside of a department is more heavily contaminated than the inside and there is no evidence that sticky mats or lines painted on the floor have any microbiological value. The same reasoning means that the transfer of patients by the two-trolley system is unnecessary (Lewis *et al.*, 1990).

The instruments and other surgical necessities exposed during an operation have an area many times greater than the average wound. Airborne microbes are more likely to get into a wound because they have fallen onto an instrument than by falling directly into the wound itself. Instruments should be exposed to the air for as short a time as possible. The instruments required for an operation may be laid out during the preceding one. This is not recommended, though if it is to be done the instruments should be prepared in conditions no less good than they will meet when they are eventually used. They must, of course, be covered in the interval.

Ventilation

The ventilation of ORs attracts a great deal of attention. There are three reasons for this: the comfort of staff and the physiological well-being of patients; the removal of flammable, explosive or toxic anaesthetic gases, and the control of airborne microbes. Although not of direct microbiological concern, it is of interest to note that the dissipation of anaesthetic gases by simple ventilation (the method still used for microbes) has proved inadequate. Special scavenging systems are now applied to the point at which they are released. The microbiological equivalent of this is to remove bacteria released from the body by enclosing it in an exhaust-ventilated, hooded suit of the type that has been worn by some orthopaedic surgeons (p.146).

The most common method for controlling airborne microbes is the plenum or turbulent ventilation system. Filtered air is introduced through vents high on the walls or in the ceiling of an OR. This mixes with the air already there. A slight positive pressure is maintained and the mixture is discharged through vents placed low down at the maximum possible distance from the inlet. As the new air enters it mixes with the old and the bacteria present are first diluted and then swept away. The incoming air is filtered to remove bacteria from it. In fact, most fresh air contains remarkably few bacteria, nearly always many fewer than are found, for example, inside an OR while it is in use. When tested years ago, fresh air in cities was found to contain spores of *Clostridium tetani* and for this and other reasons filtration became standard practice. It would be interesting to know if this is still necessary now that horses (whose droppings were almost certainly the origin of the tetanus spores) have disappeared from the streets of most cities. Sometimes the bacterially contaminated air discharged from an OR is recycled to conserve the energy used to heat or cool it. If this is done (and some authorities advise against the practice because of the anaesthetic gases) then of course it is important to filter the air before it is reused.

Another way of bringing air into an OR is by the non-turbulent laminar flow method. This introduces air through a porous area either of the ceiling immediately over the operating table or from a wall close to it, so the patient and scrubbed operating team are enveloped in a uniform, powerful, downward or horizontal flow of filtered air. This produces the ultra-clean condition mentioned in Chapter 7, p.146, and is very effective as a way of washing away bacteria released from the skin of the operating team. Perversely, however, the direction of the flow of air in the downward system ensures that any bacteria released from the upper bodies of the operating team must pass the immediate vicinity of and may be blown directly into patients' wounds *before* they are removed from harm's way. The horizontal system avoids this, provided the whole team is kept downwind of the patient.

The need for expensive high-efficiency particulate air (HEPA) filtration of the incoming air has not been explained. The same benefit might be achieved with cheaper, simpler filters. Clean room technology, introduced to keep dust out of miniaturized electronic components during assembly, has had a major impact on OR ventilation. Because all airborne particles tend to be confused with microbes (p.51), filtration has been elaborated and particles smaller than necessary are removed. This has a cost consequence. It is worth noting that there is no evidence that this level of filtration prevents surgical sepsis. Although microbes are very small, they almost always travel about on rafts that are much larger than themselves.

Air changes

The more air blown through an occupied room, the greater the dilution of bacteria released from the bodies of those within it and the lower the bacterial count achieved. At some point a balance is struck between the rates at which bacteria are added to the air by its occupants and that at which they are removed by ventilation. The flow of air can be measured in 'air changes per hour', calculated by dividing the total volume of air delivered each hour by the volume of the OR. The maximum number of changes possible is determined by the noise and draught created and the capital and revenue costs of the machinery.

Some years ago 16 changes an hour were thought adequate, now 20 or more are called for. The maximum for ordinary turbulent plenum systems may be about 30. Formerly bacteriological standards were set for ORs, based on the maximum allowable number of bacteria-carrying particles in a cubic foot or cubic metre (= 35.3 ft^3) of air. Although these numeric standards have been abandoned in favour of the physical criteria just given, they are still of interest and may even be of use in places where details of airflows are unknown. They ranged between 10 and 175 per m^3 ($<$1–5 per ft^3), depending on the country of origin and type of surgery. In this respect neurosurgery and orthopaedic surgery were the most demanding. The counts quoted were the maximum acceptable during periods of quiet operating. Counts always rise markedly at the beginning and end of surgery, when levels of physical activity are high. In an OR that has been completely empty for an hour or so the count should be zero.

Those who set out to count bacteria in the air need to be aware of two things. First, when the person who is to perform the task enters an otherwise empty space the count increases significantly. If a baseline or background reading is required the machine used must be set up some time prior to the measurement and switched on by remote control. Second, machines of several designs are available and the particles to be counted are captured in different ways. Different machines capture particles of different sizes and densities with very variable efficiencies. The result is that differences in counts may be observed when more than one type of machine is used in the same place at the same time. Some machines fail to collect a large proportion of the smaller particles present in the air and the proportion falls as the particles get smaller and lighter. It is necessary to be suspicious of machines that require the application of a multiplier to observed counts in order to arrive at the 'true' figure.

For those without the slit sampler or other device required to make direct counts of airborne bacteria, settle plates may be used. A 90 mm (3.5 inch) petri-dish of blood or nutrient agar is exposed at some convenient place near the operating table for an hour. After incubation overnight the number of colonies that develop are counted and, if required, they are identified,. To convert the number of colonies that develop after incubation into a rather crude approximation of the number of bacteria-carrying particles per m^3, the total plate count is divided by 0.13 or, for ft^3, by 4.5 (Williams *et al.*, 1966).

Rests and dirty cases

It used to be suggested that an OR should be 'rested' to allow the bacterial count to return to a low level when an operation has been completed, particularly after a 'dirty' case. In a just-emptied OR with normal plenum ventilation, the time taken for the airborne count to fall to one-twentieth of its former level is three minutes at 20 changes an hour or just under four minutes at 16. As the activity associated with the entry of the next patient will increase the count to its original level, a rest has no microbiological logic. A rest may be thought necessary because pathogenic bacteria have escaped from an infected patient or their wound. The release of a significant number of pathogens into the air is unlikely. In the case of the skin, this is because the patient is immobile and from a wound because any bacteria there are trapped in sticky liquid. Up to 50% of the population (including the staff of ORs) carry *Staphylococcus aureus* and 100% carry *Escherichia coli* or *Clostridium perfringens* as parts of their normal floras. Small numbers of these organisms, the latter the most important cause of gas gangrene, are regularly found in the air of ORs. They have come from the bodies of the OR staff, so again microbiology does not support the need for a rest. Part of the 'resting' period may be used to mop out the theatre. In the absence of soiling by body fluids this has no useful purpose. In fact, it keeps the airborne count high due to the physical exertion of the cleaner. Routine cleaning on a daily basis is all that is necessary. Disinfectants add nothing useful to the process.

Most 'dirty' cases are nothing of the kind. In this connection a smell may be thought to indicate airborne contamination by microbes. In fact, smells indicate the presence of chemicals, not bacteria, in the air. Bacteria may produce smelly chemicals, but only when they are actively growing in a nutrient environment. They cannot do it when they are suspended in the air.

ORs are not driven by microbiological considerations alone. The difficulty is that practices that may be desirable for other reasons may for convenience (or out of ignorance) be supported by spurious microbiological ideas. If the real reason for wanting to 'rest' an OR is that the staff need the rest and a cup of coffee as well, it does not help to invent a microbiological fairy-tale to justify it. Over the years those who hear such tales come to believe them. This may be one of the reasons why so many microbiological fantasies have taken root in operating departments.

INTENSIVE CARE UNITS

Patients in intensive care units suffer the highest rates of iatrogenic infections. Why this is so, what should and what need not be done about it.

Basic data

Intensive care units are hotbeds of iatrogenic infection (IaI). This is illustrated in Figure 3.3 and in Table 10.1, where more use is made of the data collected during the study in the teaching hospital (NUH) in Singapore mentioned in Chapter 3. From the table it is clear that the chance of acquiring an infection in the three major adult ICUs in the NUH was 3.5 times greater than in the rest of the hospital and that in these units 100 patients who were infected would suffer 144 IaIs with 160 pathogenic microbes. In the neonatal ICU the comparable figures were 3.4, 123 and 140. Outside ICUs, 100 patients with IaI suffered from 116 infections with 122 pathogens. Multiple infections with multiple pathogens were more common in ICUs. Compared with the risk in places where infections were least frequent ('other departments', Figure 3.3), patients in ICUs were 18 times more likely to suffer at least one attack of IaI. The reason why a patient with one IaI is more likely to acquire a second is discussed in Chapter 3, pp.29, 30 and 31.

ICUs vary enormously in their functions and in the outcomes for their patients. There are obvious differences between ICUs that care for infants, children or adults or for special groups of postoperative patients, compared with those for patients with medical conditions or

Table 10.1 The pathogens causing IaI and the rates of infection detected in adult and neonatal ICUs in the NUH, Singapore, compared with the rest of the hospital

Pathogens	% IaI, due to pathogens named, in		
	ICUs		Rest of
	Adult*	Neonatal	hospital
Escherichia coli	6	5	14
Klebsiella spp.	14	6	16
Enterobacter spp.	3	3	2
Proteus spp.	2	1	5
Pseudomonas aeruginosa	14	3	8
Acinetobacter spp.	9	2	4
All Gram-negative rods	48	20	49
ORSA	6	15	12
MARSA	17	32	11
Staphylococcus epidermidis	1	15	2
Candida spp.	13	2	4
Others	15	16	22
Infected patients/1000	90	89	26
Infections/1000 patients	130	109	30
Infecting pathogens/1000 patients	144	124	32

*Adult medical, surgical and cardiothoracic ICUs.
Key: ORSA, ordinarily resistant *Staphylococcus aureus*; MARSA, methicillin- and amino-glycoside-resistant *Staph. aureus*.

for a mixture of these, plus the victims of accidents. An additional confounding factor is the level of dependency thought to necessitate admission to an ICU. Some ICUs care for mixtures of patients who require high-dependency care as well as those who need full life support. These variables cause significant differences in the rates of infection, the types of infections and the rates of mortality in different units (Vincent *et al.*, 1995). The results of audits of outcomes conducted in different ICUs may not be comparable.

The causes of the infections

It is not difficult to find the cause of the high incidence of IaI in ICUs. Patients admitted to these units are selected precisely because they are seriously ill, so from the outset are certain to be suffering some degree of immunocompromise. Their immune competence is reduced further as life support and other devices are attached to them. One or more intra-vascular lines, an endotracheal tube and a urinary catheter are commonplace. Each of these is associated with a significant added risk of infection (p.182). Patients in ICUs are handled by staff with greater frequency and are more likely to be prescribed antimicrobial drugs than patients elsewhere. This explains why second and third infections (as well as the primary ones) are more common here than in other parts of the hospital and why these infections are more often polymicrobic.

The management of patients in ICUs is complex, requiring many special medical and nursing skills and much complicated and expensive equipment. Patients in need of these resources are concentrated in ICUs, so that the best available treatment is applied in the most cost-effective way. There is no such thing as a free lunch, however, so the benefits carry a penalty (Box 10.1). Patients who are immunocompromised are more susceptible to infection. When such patients are concentrated in ICUs they are cared for by highly trained staff, drawn from among a small pool of individuals. The staff are regularly exposed to the special hospital strains of microbes generated in ICUs so they are more likely to become temporary or permanent carriers of them (Maki, 1986). Antimicrobials are prescribed with great frequency and are often chosen from a very short list. All this

---

Box 10.1 Hard choices

Healthcare is rationed everywhere. Because of this it is necessary to examine the justification for the high cost of ICUs where, despite best efforts, mortality may exceed 25%. The money spent on each patient in an ICU would pay for the treatment of several who are less critically ill, but who are more likely to survive and return to a full, active life. This issue ought to be settled on ethical grounds. Instead, economic policies have curtailed the number of ICU beds and unfairly put the onus on individual doctors and nurses who have to refuse treatment to patients who might have benefited from their expertise.

ensures that infections, when they occur, are likely to involve similar multiresistant bacteria. The result is that debilitated patients are infected frequently and severely with organisms, often of low pathogenicity for others, that are more difficult and expensive to treat. Infections are the major contributor to the heavy morbidity and mortality experienced by patients in ICUs. Because many of the organisms responsible are of low pathogenicity they are infrequent causes of infection among healthcare workers, even when they become carriers of them.

The pathogens and where they come from

Table 10.1 displays the distribution of IaIs and of the pathogens that caused them in ICUs in the NUH, compared with their distribution in the rest of the hospital. The principal differences are that in ICUs the generally more sensitive bacteria (*Escherichia coli* and 'ordinary' *Staphylococcus aureus*) were replaced by more resistant organisms. In the NUH the replacements differ from those described in some earlier reports. This may have been due to the widespread use of the expanded range of third-generation cephalosporins. It seems that this was associated with the emergence of methicillin- (and of course cephalosporin-) resistant *Staph. aureus* (MRSA) as a major pathogen. This organism, in either its methicillin-sensitive or -resistant form, also achieved the dominance as a pathogen expected of it in neonatal ICUs (p.160 and 161).

The scale of the problems listed make the prevention of infections in ICUs an uphill task. The best that can be achieved is to reduce their number to some as yet undefined minimum. Many are self-infections of the autogenous sort and some are cross-infections. Other than in burns units (p.224), environmental infections are only important if there is an accident or major carelessness. These usually involve the wet environment, particularly apparatus that contains liquid or collects it in the form of condensation. This generalization breaks down in the case of some transplant patients who are unusually susceptible to fungal infections, particularly with *Aspergillus* spp., the spores of which are transmitted through the air. Other than as just noted, however, there is little evidence that the dry environment or the air contribute significantly to the rate of infection in ICUs; indeed, the opposite is the case (Bauer *et al.*, 1990).

Protective clothing

Such things as overshoes, special headgear or adhesive ('sticky') mats at the entrance are not required in ICUs. Masks and gowns (Haque and Chagla, 1989; p.195) contribute little or nothing to the control of infection and special disinfection is not necessary. The part played by staff clothing in the transfer of infection is difficult to assess, which probably means that it is not very important. To reduce the risk and to protect clothing from direct contact with heavily contaminated sites or splashes, plastic aprons may be used when performing such procedures as endotracheal suction. These provide more protection than fabric gowns, though if satisfactory gowns made of waterproofed or other less pervious materials were available, they would be preferable. This is more important if the task involves a risk of significant contamination of arms or shoulders left unprotected by the type of plastic apron most often supplied.

Hands again

In ICUs, as everywhere, the hands of staff are believed to be responsible for most transfers of bacteria that colonize or infect patients (p.178). Hands commonly acquire bacterial pathogens as temporary residents by contact with patients who are already colonized or infected. A bacterial pathogen that is part of the resident flora of the hands of a staff member is of disproportionately major significance (Maki, 1986). In either case hospital bacteria are transferred with great efficiency to colonize and infect new patients. It is commonly thought that this route can be broken if each patient is tended by staff on an individual basis. This may be attempted by providing a separate nurse for each patient, but the attempt breaks down because it cannot be achieved for doctors, physiotherapists, respiratory and X-ray technicians, venepuncturists and the like. Even for nurses the arrangement may fail at night. The situation is much worse if ICUs are crowded or understaffed or if handwashing facilities are inadequate.

The methods that ought to be used to prevent IaIs in ICUs are, of course, exactly the same as those that should be used everywhere. The difference lies in the exceptional sensitivity to infection of the patients concerned, compounded by the intensity of their exposure to the risk of it. The result is that in ICUs small breaks in technique that might not cause harm elsewhere are not forgiven. The documented failure of staff to wash their hands (p.178) means that little will be gained by looking elsewhere for the causes of infections in ICUs; indeed, the expensive pursuit of imaginary problems might develop into a cult, as has happened in operating departments. If it were possible to impose proper hand hygiene this would have a dramatic effect, but it would not be the end of the matter. If hands were washed properly at the proper frequency this would consume so much time as to require additional staffing. It is also probable that sore hands would result, arguably producing an even greater problem with infection. As has been noted, this is an area that calls for new thinking.

BURNS UNITS

A burns unit is like an intensive care unit that keeps seriously compromised patients for longer, so problems with infections are compounded.

There are two reasons why burns wounds are particularly susceptible to infection. First, a burn provides an area where water and nutrients are plentiful, ideal for bacterial growth. Second, in a burn the body's normal defences against bacterial invasion are weakened or absent. Not only is the protective epidermis destroyed, but the initial coagulum and later a slough form a barrier to the entry of phagocytic cells, antibodies and the other components of the immune system. Colonization of the surface of a burn is thus inevitable. This may be initiated by microbes that, due to drying of their surfaces, have lost the power to infect normal tissues (pp.42 and 137). The body may then fail to react to prevent an abnormal colonisation from becoming an infection. Although it is difficult to determine clinically the point at which the change takes place, its eventual effect is all too obvious. Patients with extensive burns who survive the initial injury suffer a significant mortality later, mostly due to infections.

Lives are saved by aggressive resuscitation as soon as possible after a burn. Proper care will prevent the injury from extending unnecessarily and in the end skin grafting is often required. The skills that contribute to this process are best applied in burns units, which are specialist intensive care units that keep their patients for longer than usual. Practically everything said about ICUs in the previous section also applies to burns units but the longer stay makes the control of infection even more difficult.

Burns wounds are not only uniquely receptive of microbes but when colonized or infected, they are also prolific sources of bacteria. Sheer weight of numbers encourages the transmission of infection by routes of little or no importance elsewhere (Ransjo and Hambraeus, 1982). Of course, hands are still a principal route of transmission, but now clothing, the environment and the air can all play a part. The more prolonged treatment of burns patients ensures that the larger number of routes along which microbes can spread also remain open for longer. Even if transmission by the hands were controlled, infections would still spread to cause problems. Conversely, because burns are so different, techniques used to control infections in these units should not be applied uncritically to other areas. This has happened in the past, with expensive consequences (Colebrook, 1948; Box 7.4; p.135).

At any one time the bacteria that cause infections in a given burns unit tend to be indistinguishable. The cause of this is as described for ICUs but for the reasons just given, they apply with greater force in burns units. *Staphylococcus aureus* (including MRSA) and a variety of Gram-negative organisms, chiefly *Pseudomonas aeruginosa*, are the most common pathogens. Formerly the more pathogenic *Streptococcus pyogenes* was also common, but fortunately (as this coccus is also fatal to skin grafts) it is now less usual. Each unit seems to provide a special niche that favours one or other of the major pathogens or perhaps a group of them that come and go over periods of months or years. The conditions in different units that favour one pathogen over another have not been defined, though they deserve study.

It is generally agreed that infection is inevitable in burns that involve more than about 30% of the body surface. This fact has led to the practice, whenever possible, of early radical excision of some or all of the burn, with closure by grafting (Demling, 1983). In the meantime control depends on the judicious use of antimicrobials and the application of the precautions described for ICUs, noting the importance in transmission of the arms as well as the hands of staff (Burnett and Norman, 1990). To these may be added filtered ventilation (with attention to the direction of the airflow) and special clothing. A burns unit sited in a general hospital may become an important permanent reservoir of pathogens that can spread to other patients. The physical segregation of burns units should be as complete as possible and the staff working in them should not be transferred thoughtlessly to other areas, particularly to ICUs or oncology units.

OTHER SPECIAL DEPARTMENTS AND UNITS

As another intensive care unit with a special function, renal units were the first to suffer acutely from the effects of bloodborne viruses. The lessons learnt have a general applicability.

Renal units

In the last 50 years dialysis and transplantation have completely revolutionized the treatment of end-stage renal failure. Remarkable advances have been made in the design of dialysers, in the techniques of arteriovenous access, by the introduction of peritoneal dialysis and not least in the development of a high level of skill in their application. Fundamental advances have also been made in the field of transplantation. These are highly specialized areas, with their own voluminous literatures. This is required reading for those who work in them and it would be futile to attempt to summarize the infection control element of this literature here.

For the general reader, however, there are lessons to be learnt from the experience gained in the management of iatrogenic infections in renal units. In their early days outbreaks of serious infections among both patients and staff threatened not only the future of many units, but possibly the whole of the programme. The problems were exceptionally acute, conspicuous and sufficiently novel for them to be accompanied by strong negative reactions among healthcare workers. Similar reactions accompanied the emergence of the HIV virus in the 1980s and have reappeared in connection with multiply resistant *Mycobacterium tuberculosis* in the 1990s. In the case of renal units the problems were not unique when taken individually, but their frequency and high-profile severity were. The control measures developed are applicable in other clinical areas.

The first difficulty was that the apparatus used to treat end-stage renal failure was (and still is) expensive and the skills needed to use it expensively acquired. Treatment is most effectively delivered in or from specialist units where patients of the same type are concentrated and exposed to essentially identical regimes. In this respect renal units resemble intensive care and burns units (pp.220 and 224) except that their patients are exposed to the risk of infection for very much longer. The second problem is that the patients are somewhat immunocompromised by their underlying disease, so are more easily infected and their infections are more difficult to treat. Third, for haemodialysis, direct access to the cardiovascular system is required, repetitively, for significant periods of time on each occasion. This exposes both patients and staff to the risk of bloodborne infections and the patients to infections at the site of vascular access. These can develop into septicaemias. In the case of peritoneal dialysis, similar regular access is required and any infections that develop at the site of catheterization soon spread to cause peritonitis. Fourth, patients with chronic renal failure become anaemic and require occasional blood transfusions. The use of haemopoietin has reduced but not eliminated the need for these, which once more expose patients (and so staff) to the risk of bloodborne infections.

In the early days of renal units haemodialysis was the norm, significant spillage of blood a daily occurrence and machinery, including the dialysers, was disinfected, reused and shared between patients. There were no laboratory tests for the detection of bloodborne viruses. In these circumstances a patient who carried one of these viruses might be admitted to a renal unit, where it would spread to other patients and because the hepatitis B virus (HBV) is highly infectious, to members of staff as well. The introduction of tests for hepatitis B surface antigen (HBsAg) and later for others of its viral components made a great difference. It was now possible to identify carriers of HBV among patients prior to admission to a unit, to test members of staff and to exclude any blood donors who, as carriers, could transmit the virus. HBsAg-positive patients were refused admission to some units or they were offered treatment in special isolation facilities. Although it is much less infectious, the same precautions were applied as soon as a test for carriage of HIV became available and the process is now being repeated with the hepatitis C virus (HCV). The availability of an HBV vaccine was another positive step in the protection of members of staff although it is less useful for patients, many of whom, because of their immunocompromise, fail to respond to it. An anti-HBV hyperimmune human immunoglobulin is available for the protection of those exposed to the possibility of HBV transmission as the result of an accident.

One of the measures that served to reduce the possibility of transmission of these viruses between patients and between patients and staff was the introduction of home dialysis. Routine treatments were managed by the patients or their relatives, in their own homes. This movement was accelerated by a catheter that made access to the peritoneal cavity more easily managed and continuous ambulatory peritoneal dialysis (CAPD) became fashionable. Of course, barrier precautions that protect the eyes, nose and mouth are taken and gloves are worn for all patient contacts when exposure to blood is possible. The introduction of more compact and eventually disposable dialysers has added an extra safety factor and even if these are reused, each unit is reserved for a single patient. The non-disposable parts of dialysis machines may also be kept for single patients or shared between the same small groups of them. Rules about blood spills and the disposal of 'sharps' and other waste are strictly applied and each unit has a safety officer who records accidents and monitors compliance with the rules. In units managed in this way HBV infection has ceased to be a problem and HIV infection never has been one. The final position with HCV has yet to emerge, but the same precautions serve for all three viruses.

The insertion sites of the plastic catheters used to secure access to the cardiovascular system or to the peritoneal cavity suffer the difficulties encountered by all such devices that combine a unique point of entry, a wound and the permanent presence of a foreign body (p.182). Because the number of possible alternative sites is limited, each entry site must be preserved for as long as possible. Elaborate protocols have been drawn up and applied to delay the development of infection and to limit its extent. Renal units have all the attributes required to generate a problem with multiply resistant bacteria and *Staphylococcus epidermidis* and *Enterococcus* spp., in particular, have duly obliged. Peritonitis or septicaemia with methicillin- or vancomycin-resistant strains of these bacteria are not easily treated.

Other types of unit

Space does not permit individual consideration of the particular problems of the growing range of small units that have grown up round groups of patients with special needs or round complex and expensive types of apparatus that serve specialist diagnostic or therapeutic purposes. Examples are the invasive techniques involved in imaging in all its old and new varieties, endoscopy, cardiopulmonary and urological investigations, laser treatments and colposcopy. The characteristics of the patients involved, the nature of the interventions made and the type of equipment used determine the scale and types of IaI that are likely to be encountered in each case. The identification of these will suggest the nature of the control measures that are appropriate.

The people who set up and work in these special units are nearly all enthusiasts to whom infections are side issues that tend to be ignored until they are endemic and it is too late to do anything effective about them. Such units need to be kept under review by infection control organizations, particularly in the early stages of their development when novel equipment is unfamiliar and there is a tendency to allow the service to grow beyond what the facilities can safely accommodate.

WET ENVIRONMENTS

Water is essential to life. Most pathogenic microbes die more or less quickly when deprived of it. In healthcare premises dry environments are safe; wet ones may not be.

Life is impossible without water. Most living things die rapidly if deprived of it though some can 'hibernate', if necessary for years, until water returns. In the absence of water microbes immediately cease to multiply and most begin to die more or less quickly. Some last longer than others, but excepting the bacteria that form spores or have developed some other special means of protection against desiccation, few survive longer than a period measured in minutes or hours. Otherwise sensitive microbes may be protected from lethal desiccation if, for example, they are embedded in material that contains proteins or their breakdown products.

Microbes cause infections only if they are present in significant numbers and can reach their victims. A totally dry environment that is also clean is, with great certainty, a microbially safe environment. If no new microbes are added it becomes safer as time passes. A wet environment is exactly the opposite. Unless special precautions are taken wet environments may be microbially dangerous at the outset and are likely to become increasingly so with time. Wet environments, however, are only potential hazards. There must be a connection (a vector or vehicle) to convey any microbes they contain to a portal of entry on a potential host, otherwise they cannot initiate an infection (p.42). On this basis things that are wet can be identified immediately as being potentially dangerous and they can be distinguished very clearly from dry items that are unlikely to cause a problem. Wet environments in healthcare facilities include intravenous and other liquid medications, dialysis, irrigation and rinsing fluids, medical equipment that contains liquids or collects them in the form of condensation, disinfecting and antiseptic solutions, food and drink, all plumbing, wet cleaning equipment, waste and sewerage systems, parts of air-conditioning plants and hydrotherapy pools. To complete the list, wet linen, wet bedpans, urinals and bowls, surfaces left wet after cleaning and ornamental water and flower vases must be added. Unless a patient drinks the water from a flower vase, however, or it is used to irrigate a wound, this is only a reservoir and not a source of infection (p.43).

With variations, bacteria or fungi of one kind or another can multiply in most of the fluids found in hospitals, from distilled water to solutions of disinfectants. The more pathogenic types of bacteria generally grow poorly or may fail to survive in less nutrient liquids.

Weak pathogens or normally non-pathogenic free-living bacteria, on the other hand, may multiply to reach very large populations. Such liquids are often inherently poisonous even if the microbes concerned cannot multiply in living tissues to cause infections, so are non-pathogens. Fluids in this state are normally kept away from critical areas by intact body surfaces, but medical procedures open routes through these surfaces to allow liquids to reach areas otherwise inaccessible to them. Debilitated hospital patients are specially vulnerable to insults of this kind.

The only fluids that can be guaranteed as being microbiologically safe are those that have been prepared with great care to exclude microbes at every stage of manufacture and have been sterilized properly (usually in an autoclave) inside a hermetically sealed final container that remains intact. Once the container has been damaged or opened the fluid is compromised and the chances that it will remain sterile diminish progressively as time passes. The importance of this varies with the purpose to which the fluid is to be put. An intravenous infusion, for example, requires more elaborate care than a can of soft drink, but the principles are the same. As is the case with food, refrigeration slows down and freezing halts the multiplication of bacteria and fungi. Viruses require the presence of liquid water inside the living cells in which they multiply, so in the context of the environment they are not a problem.

Modern plumbing and the wet parts of air-conditioning systems are happy hunting grounds for whole armies of mostly non- or weakly pathogenic microbes. It should be remembered that even high quality drinking water is not sterile. Hot water may not be hot enough for long enough to disinfect itself and the cooler parts of hot water systems can act as bacteriological incubators. Depending on the quality of the supply and the design and state of their plumbing systems, the water in healthcare facilities (and elsewhere) may contain large numbers of bacteria. These usually pose no threat to healthy members of staff but can be dangerous to patients. They may include *Legionella pneumophila*, the cause of legionellosis. The air delivery ducts of air-conditioning systems are (or should be) dry, so air does not pick up microbes as it passes through them, but this may not be true if there is water in the system. Air-conditioning cooling systems and humidifiers may be wet, so are suspect. The dangers can be reduced by good design and regular skilled maintenance.

The wet environment is controlled by limiting its extent as far as possible and by applying sensible controls to what is left. These include keeping sterile fluids intended for parenteral use sealed in their containers until the last possible moment, taking great care when making additions of medications to infusions and limiting the time any delivery system is kept attached to a patient. Nasogastric feeds used for enteral hyperalimentation have been identified as an important source of iatrogenic infections (Thurn *et al.*, 1990). These feeds should be prepared in scrupulously clean and disinfected equipment. They should be kept refrigerated until needed and should not be stored for more than about eight hours. Remember, when a fluid is being run into a patient it is exposed to room temperature. If it is or becomes contaminated, bacterial multiplication will continue at a compound rate for however long the process takes to complete.

There is no place for regular, routine microbiological testing of the environment, wet or dry, with the possible exception of infant feeds if these are prepared in the hospital, of piped water if the quality of the supply is uncertain or variable or of water disinfected by ultraviolet light. Microbiological testing *is* required when a problem arises and it should be directed at its most probable cause. Testing should cease when the cause of the problem has been located and a solution has been found, has been applied and has worked. Remember that an apparent source of microbes may turn out to be another victim of the real source, still undiscovered.

REFERENCES

Bauer, T. M., Ofner, E., Just, H. M. *et al.* (1990) An epidemiological study assessing the relative importance of airborne and direct contact transmission of microorganisms in a medical intensive care unit. *Journal of Hospital Infection*, **15**, 301–9.

Bethune, D. W., Blowers, R., Parker, M. *et al.* (1965) Dispersal of *Staphylococcus aureus* by patients and surgical staff. *Lancet*, **i**, 480–3.

Burnett, I. A. and Norman, P. (1990) *Streptococcus pyogenes*: an outbreak on a burns unit. *Journal of Hospital Infection*, **15**, 173–6.

Colebrook, L. (1948) The control of infection in burns. *Lancet*, **i**, 893–9.

Dankert, J., Zijlstra, J. B. and Lubberding, H. (1979) A garment for use in the operating theatre: the effect upon bacterial shedding. *Journal of Hygiene, Cambridge*, **82**, 7–14.

Demling, R. H. (1983) Improved survival after massive burns. *Journal of Trauma*, **23**, 179–84.

Haque, K. N. and Chagla, A. H. (1989) Do gowns prevent infection in neonatal intensive care units? *Journal of Hospital Infection*, **14**, 159–62.

Laufman, H., Montefusco, C., Siegal, J. D. *et al.* (1980) Scanning electron microscopy of moist bacterial strike-through of surgical materials. *Surgery, Gynecology and Obstetrics*, **150**, 165–70.

Lewis, D. A., Weymont, G., Nokes, C. M. *et al.* (1990) A bacteriological study of the effect on the environment of using a one- or two-trolley system in theatre. *Journal of Hospital Infection*, **15**, 35–53.

Maki, D. G. (1986) Skin as a source of nosocomial infection: directions for future research. *Infection Control*, **7** (supplement), 113–16.

May, K. R. and Pomeroy, N. P. (1973) Bacterial dispersion from the body surface, in *Airborne Transmission and Airborne Infection*, (eds J. F. Ph. Hers and K. C. Winkler), Oosthoek, Utrecht, pp. 426–34.

Meers, P. D. and Yeo, G. A. (1978) Shedding of bacteria and skin squames after handwashing. *Journal of Hygiene, Cambridge*, **81**, 99–105.

Ransjo, U. and Hambraeus, A. (1982) When to wash walls in ward rooms? *Journal of Hospital Infection*, **3**, 81–6.

Speers, R., Bernard, H., O'Grady, F. *et al.* (1965) Increased dispersal of skin bacteria into the air after shower-baths. *Lancet*, i, 478–80.

Thurn, J., Crossley, K., Gerdts, A. *et al.* (1990) Enteral hyperalimentation as a source of nosocomial infection. *Journal of Hospital Infection*, **15**, 203–17.

Vincent, J-L., Bihari, D. J., Surer, P. M. *et al.* (1995) The prevalence of nosocomial infections in intensive care units in Europe. *Journal of the American Medical Association*, **274**, 639–44.

Williams, R. E. O., Blowers, R., Garrod, L. P. *et al.* (1966) *Hospital Infection, Causes and Prevention*, 2nd edn, Lloyd-Luke, London, p. 372.

The role of service departments in infection control

OCCUPATIONAL HEALTH, STAFF VACCINATION AND
SAFETY

**Occupational infections are of particular concern to healthcare workers.
This section defines the extent to which the concern is justified and says
what may be done about it.**

Healthcare for healthcare workers

It is unacceptable for an industry concerned with the delivery of
healthcare to ignore the health of its own employees. Even when it is
not a legal obligation, employers should make a point of caring for
the well-being of those who work for them. Regrettably, this does not
always happen.

Many hospitals and other healthcare facilities operate comprehensive staff health programmes, providing medical examination on entry,
immunization and a 'walk-in' service for medical problems (including
injuries) that arise at work. Infections and the fear of them rank high
among the concerns of healthcare workers (HCWs). An important
part of the duties of an occupational health service is to monitor infections that might have a temporary or permanent effect on employabil-
ity. It should also prepare detailed policies and supervise their
application to the management of HCWs who have been exposed to
the risk of infection, particularly with the bloodborne viruses. The
service should include counselling. Infection control teams (ICTs) need
to maintain very close links with such organizations and infection
control committees (ICCs) should contribute to the production of
guidelines for the immunization of HCWs.

Apart from the general responsibilities of employers, the existence of
infections in healthcare premises adds a major additional component
to the duty of care borne by healthcare managers. An HCW with
open tuberculosis can infect dozens of patients and one with mild
chickenpox or shingles can be the source of a lethal infection in a
child under treatment for leukaemia. HCWs who carry one of the

bloodborne viruses, in particular hepatitis B and C and to a much lesser extent the human immunodeficiency virus (HIV), may pass these infections on to their patients. Equally, an HCW may contract hepatitis B or C or HIV infection from a patient or a pregnant HCW may lose her foetus or give birth to a deformed infant after contact with a case of rubella. Managements should seek to prevent accidents like these by setting up and supporting organizations of the kind outlined above.

Education is a vital part of any staff healthcare programme. Introductory courses for different groups of HCWs on first employment should include instruction on how to avoid infecting themselves or the patients they serve. The importance of reporting all accidents should be stressed. Insistence on a 'safety first' attitude to work is part of the process. This will range from the right way to lift objects so as to avoid back injuries to the proper handling of sharp instruments to avoid infections (p.202). These messages should be reinforced by repetitive in-service education.

Integral to a staff health programme is an immunization policy. This should be designed to protect HCWs from infections they might contract from patients and to protect patients from infections they might acquire from HCWs. Requirements will vary according to the health problems in the community served by the hospital and according to the categories of staff. A review would include tuberculosis (which also requires a decision on the need for routine chest X-rays of staff, at least on first appointment), rubella, poliomyelitis, diphtheria, tetanus, hepatitis B and varicella. Other vaccines that might be considered in special circumstances are those for meningococcal meningitis, *Haemophilus influenzae*, hepatitis A, typhoid, cholera and yellow fever, plus others in less usual conditions. In the case of varicella, if the vaccine is not to be used, the supposed or actual immune status of HCWs needs to be recorded to aid in the allocation of staff to high-risk areas.

Infections in healthcare workers

By comparison with many other industries, HCWs are employed in a very safe environment. Despite this many HCWs believe they are exposed to an increased risk of infection. Experience has shown that they are right, but that the risk is small. No matter how small, however, employers and employees share a duty to minimize it. In several countries legislation now makes this a statutory obligation. Employers must document the precautions introduced to prevent infections in their employees and employees are required to apply them.

The hazard

Infection is a natural and so far unavoidable consequence of communal life. The risk of acquiring an infection is accentuated in hospitals and other healthcare establishments. To put this in perspective, most individuals suffer one or more minor infections each year. Sometimes these are more severe and more frequent, particularly in childhood and old age. Apart from hepatitis B, now controlled, few HCWs who have spent a lifetime in the industry have acquired a serious infection as a result of their employment. The more common minor infections are a hazard in any industry where employees have regular, face-to-face contacts with members of the public.

HCWs cannot fail to notice that a significant proportion of the patients they meet are under treatment for an infection. Many of these infections are complications of the illnesses for which they have come for treatment or are unwanted consequences of that treatment. Very few of the microbes responsible are a threat to healthy HCWs or healthy members of their families. An example is the MRSA. Between 25% and 50% of all individuals carry 'ordinary' more sensitive *Staph. aureus* as a part of their normal flora, nearly always without knowing it. HCWs may acquire MRSA from patients who carry or are infected with it. MRSA is more resistant, but it is not more pathogenic than other *Staph. aureus* (p.165). The HCW has the same small risk of developing an infection no matter which strain of staphylococcus they carry (Box 11.1). The only danger is that they might transmit their MRSA to the next patient they care for. As with any staphylococcus, this might cause an infection but, in the case of MRSA, one that is a little more expensive to treat. In these circumstances some have thought it expedient to exclude the carrier from work until they are rid of their MRSA. Nevertheless, this microbe falls into the negligible category of risk for HCWs.

Not all microbes belong to this low-risk category. Infections caused by more dangerous microbes fall into two groups. In the first are those that threaten people (patients or staff) who are pregnant. Examples are chickenpox, shingles and rubella. HCWs may be

---

Box 11.1 The demonization of MRSA

Staphylococcus aureus is inseparable from the human race because, for many, it forms part of their normal bacterial flora. It rarely causes infections in healthy individuals. It is more commonly found as a pathogen in the debilitated but even then most of the infections it causes are mild. Some strains of *Staph. aureus* are more pathogenic than others, particularly if they carry the extra genetic information that allows them to manufacture the powerful toxins that cause food poisoning or the scalded skin or toxic shock syndromes. There is no connection between antimicrobial resistance and pathogenicity. A staphylococcus that has acquired the genetic ability to resist methicillin is not any more or any less pathogenic than its methicillin-sensitive forebear. Regrettably, for reasons that are complex, when *Staph. aureus* acquired methicillin resistance, some saw the change as an example of demonic possession and resurrected the rituals formerly employed against *Streptococcus pyogenes* in what, it now begins to appear, was an expensive and fruitless attempt at exorcism.

protected from these as a result of prior natural infections or by vaccination or in some cases by the use of hyperimmune globulin after exposure. Unprotected pregnant HCWs should not be required to perform duties likely to bring them into contact with these infections.

The second group of microbes are those that can attack healthy people, with serious consequences. Infections with the hepatitis B virus (HBV), the human immunodeficiency virus (HIV) and tuberculosis (p.172 and 173) fall into this class and are the subject of widespread concern. The hepatitis C virus (HCV) may join them (Heptonstall and Mortimer, 1995).

Bloodborne viruses

HBV, HCV and HIV are spread by intimate activities that involve the transmission of body fluids, blood in particular, from one person to another. In the community they are spread almost exclusively by sexual intercourse, intravenous drug abuse or vertically, from an infected mother to her infant. When therapy is involved procedures that range from contact with body fluids through the use of hypodermic needles to blood transfusion and major surgery offer artificial, highly effective routes for the transmission of HBV, HCV and, to a lesser extent, HIV. The reason for the difference is that a given volume of the blood of an HBV carrier contains perhaps hundreds of time more infectious doses than is the case with a carrier of HIV. Immunization provides a high level of protection against hepatitis B. Every HCW should be vaccinated against this disease and it should be an obligation upon management, through their occupational health departments, to see that this is done. There is as yet no effective vaccine against HCV and HIV. An HIV infection curtails lifespan and can put the sufferer at a disadvantage in respect of such things as life insurance and mortgages. It is greatly feared. For these reasons this infection is considered separately, below.

The problem with HBV, HCV and HIV infections is that many individuals are reluctant to admit or they may be unaware that they are carriers of one of these deadly or potentially deadly viruses. The carrier status can only be revealed by laboratory tests. The relevant tests may not be performed on patients under treatment for conditions unrelated to these infections and there has been resistance (partly for the reasons noted above) to the suggestion that all patients and staff should undergo these tests as a routine. Another reason for reluctance is that the performance of a large number of tests inevitably throws up a proportion of false-positive results. In areas of low prevalence of an infection these can equal or outnumber true positives (Box 11.2). A further reason is that testing may breed a false sense of security. HCWs ought to plan their activities in the knowledge that some of their patients (and even their colleagues) may be carriers of these viruses. If 'everyone' were tested some would escape the net. Evasion

---

Box 11.2 Laboratory tests are not infallible

For a variety of reasons, including human error, no laboratory test is ever 100% accurate ('specific'). The better the test the smaller the margin of error, but some negative results will always emerge as positive and vice versa. A reasonably accurate test might give correct answers on 99.5% of occasions. This is acceptable in individual cases, provided the result coincides with other evidence, for example, the nature of a patient's symptoms. It is less acceptable if the test is used to screen healthy people or patients admitted for treatment of a condition unrelated to the test in question. Many authorities have concluded that it is unacceptable when, as is the case with the human immuno-deficiency virus (HIV), a positive result is not only a sentence of death but also carries other disadvantages.

The reason is that if 1000 HIV tests are done with (say) 99.5% accuracy, five false results will emerge. If these are falsely positive, then five individuals are wrongly diagnosed as being infected with the HIV virus. In a community with a low rate of HIV infection, say two in 1000, the false-positive rate is *higher* than the true rate, though very few who are really infected will be wrongly reported as negative. This is why specimens found to be HIV positive are retested by a different method. The chance that both tests simultaneously give false-positive results is very small indeed.

---

is more likely among those who fear or who already know but have concealed a positive result.

Additional complications are what have been called 'windows of uncertainty'. There is always a delay between requesting a test, its performance and the notification of a result, particularly if it is positive and must be confirmed by a second test. This may take some days after patients have come under care and the appropriate specimens and requests have been submitted. The most dangerous period for HCWs is precisely this early period when most invasive diagnostic tests and therapeutic procedures are performed. A second 'window' follows immediately after infection when the individuals concerned may be able to transmit the infection to others, but have not yet developed the tell-tale antibodies upon which laboratory diagnosis usually depends. The latter window is also of concern in blood transfusion and transplantation.

At present many HBV, HCV and HIV carriers remain undetected as they pass through healthcare facilities. This is likely to continue. Healthcare workers must approach every patient as if they might be infected with one of these agents. Simple barriers, gloves for example, are used when exposure to body fluids is expected (pp.188 and 195).

There is less controversy about the other precautions that are necessary. These relate to the way in which clinical waste is handled (p.200). The most important part of this is to reduce the number of accidents that result from the use and disposal of sharp diagnostic or therapeutic tools. Accidental cuts and punctures from sharp objects contaminated with the blood of patients are common. These may involve the primary users of these items or those who handle the containers of waste in which they are transferred to their final resting places. Although the attempt to prevent these injuries is not controversial, the methods that may be adopted to achieve it are. The reasons for this are very practical and they include genuine differences of opinion and the shortage of money. Again, local decisions must be taken in the light of local circumstances.

The other type of precaution that may be taken is to inform all HCWs who may be involved when a patient known or thought to have a dangerous infection comes under care. They can then take any additional protective measures thought necessary. This approach may be modified by local rules concerning the confidentiality of this kind of information and procedures of this sort must be defined before it is necessary to use them. It is stressed that the existence of such arrangements must not be allowed to form another basis for a false sense of security, to conceal the fact that most patients with these potentially dangerous infections will escape any special precautions because they are not identified in time, if at all.

Occasionally an HCW is discovered to be carrying HBV or HIV. If their duties include diagnostic or therapeutic activities that might lead to the infection of patients, these must cease. Infected surgeons and others have been responsible for several transmissions of HBV to patients as a result of accidental cuts of fingers and punctures of gloves during invasive procedures. The transmission of HIV in this way is less likely because of the larger volume of blood that must be transferred and extremely few patients are recorded as having been infected by HCWs. Decisions in these matters may involve the future careers of expensively trained individuals so they are not taken lightly, but the peace of mind of a nervous public cannot be ignored. Many health authorities or regulatory bodies have made recommendations or regulations that govern practice in this difficult area.

Occupational HIV infections

Before the first cases of AIDS were recognized in 1981 its cause, HIV, had already spread widely beyond its primary focus in Sub-Saharan Africa. In the last 20 years the pandemic has grown to the point that WHO predicts 30–40 million cases of HIV infection by the year 2000, with deaths from AIDS at about a million a year (two every minute). The first recorded case of occupational HIV infection was reported in 1984. By early 1995 the number of cases had risen to 214 (Monthly Report, 1995; Working Party, 1995; Tables 11.1 and 11.2). A glance at Table 11.1 shows that these figures are very incomplete. It is inconceivable, for example, that the USA has borne the brunt of these infections. Lack of money to perform the tests necessary for a diagnosis, coupled with what have sometimes been discreditable religious, political and economic motives, underlie a gross under-reporting of HIV infection and AIDS from many countries. However, in the context of occupational infections the figures are of great importance and give some comfort. The figures from the USA and Europe represent the result of many billions of separate, more or less intimate contacts between healthcare workers and their patients, among whom a not inconsiderable number of the latter were, at the time, infected with the HIV.

Although they are useful, the problem with much of the data in Tables 11.1 and 11.2 is that they were collected retrospectively. Even the one-third of cases of infections recorded as 'certainly occupational' include many HCWs who were not tested for their HIV status at the time of the accident thought to be responsible. It is probable that some at least failed to reveal other risk factors and had in fact acquired their HIV infections before the alleged or actual incident. This means that even these tiny figures overestimate the true level of risk.

Much better data come from prospective studies in which details of incidents in which HCWs were exposed to the risk of infection from known HIV-positive patients are recorded very soon after an accident has taken place. The HCWs are tested and those whose immediate

Table 11.1 Cases of human immunodeficiency virus (HIV) infection among healthcare workers (HCWs), judged to have been acquired occupationally (figures from Working Party, 1995; see text and Table 11.2)

Exposure	USA	Europe	Rest of World	Totals
Certain[1]	43	25	5	73
Possible[2]	91	41	9	141
Totals	134	66	14	214

Key: [1] Certain: exposure documented, with an accident involving an HIV-positive patient, most due to percutaneous injuries with a sharp object, in which the HCW was subsequently shown to be HIV positive.
[2] Possible: proven HIV infection in an HCW but no accident was documented. The HCW concerned denied other risk factors for HIV and independent assessment judged the infection to have been occupational.

Table 11.2 Categories of the HCWs included in Table 11.1, who were thought to have been occupationally infected with the HIV virus

Category of HCW	Certain	Possible	Totals
Nurses[1]	35	50	85
Technicians[2]	24	28	52
Doctors & dentists	7	33	40
Other HCWs	7	30	37
Totals	73	141	214

Key: [1] Includes midwives, nursing aids, etc. [2] Includes laboratory workers, surgical, dialysis and respiratory technicians, etc. and paramedics.

postexposure tests are negative are tested again at intervals for HIV seroconversion. It is accepted that if these tests are still negative six months after the accident the HCW has not been infected. Many studies of this sort are in progress. By mid 1995 they had recorded 6325 separate instances of exposure of seronegative HCWs to infected patients, with full postexposure serotesting (Working Party, 1995). Twenty occupational infections were detected, an incidence of one infection for every 320 exposures to known HIV-positive fluids in which skin was punctured. In addition, there were fewer than one infection for every 1000 mucocutaneous exposures to HIV-positive fluids recorded. A more detailed study of the procedures leading to accidents that resulted in HIV infections revealed that 75% were associated with apparatus used for arteriovenous vascular access, outstandingly venepunctures. This is accounted for by the greater volume of blood that can be inoculated as a result of a puncture by a hollow needle compared with a solid sharp object such as a suture needle or scalpel blade.

These low levels of risk mean that it is quite difficult for HCWs to acquire HIV infections occupationally. When it is recalled that the exposures were all considered as accidents and that accidents are avoidable, the fear of occupational HIV infection falls into its correct perspective. This must not encourage complacency; the price of safety is constant vigilance.

CENTRAL STERILE SUPPLY

Central sterile supply organizations provide a specialist service. Their functions are outlined.

Introduction

Lister introduced antiseptic surgery over a short period towards the end of the 1860s. Its replacement, the unfortunately named aseptic surgery (p.215), emerged piecemeal at different times in several countries and it continues to develop. From the outset, however, one major requirement separated aseptic from antiseptic surgery. This was the need to sterilize surgical paraphernalia. In 1886, in Germany, the steam sterilizer or autoclave was adapted to achieve this.

What happened next varied in different parts of the world. In many places boiling water was used to make metal surgical equipment safe, while fabrics – surgical dressings, gowns, drapes and so on – were autoclaved. These practices spread from operating departments into wards and clinics, where the separate use of so-called boiling sterilizers for metal items and of autoclaves for dressings continued. Each ward and clinic had its own boiler, tended by nurses. The large complicated autoclaves were kept out of sight. The early machines with their masses of exposed pipes, valves and dials were given to making loud noises and to the violent discharge of quantities of hot water and steam. It was difficult to find sufficiently knowledgeable people to take responsibility for these rather fearsome objects and the duty often fell to junior persons with strong nerves and limited imaginations. In these circumstances autoclaves were often badly misused.

These arrangements were clearly unsatisfactory. Sterilization to modern standards was unusual and some equipment was dangerous to users. It was fortunate that, due to the massive overkill inherent in an autoclave (p.108), not too many patients suffered accidents as a result of failures to sterilize. More notable was the inadequacy of management and the lack of supervision. To these deficiencies was added the low cost-effectiveness of using ward nurses to prepare, one by one, each of the settings needed for the enormous number of 'sterile' procedures carried out every day. A central service was required that could perform the whole function, to produce and distribute standard securely wrapped sterile packs each containing what was necessary for a given procedure. In the 1930s action along these lines began to be taken in the USA and in the 1940s the idea and the practice spread more widely, partly as a result of military need.

The development of central services created a demand for a whole range of new machinery. This comprised more efficient high and low temperature autoclaves, gas sterilizers, sophisticated apparatus for washing and cleaning instruments and other appliances. The appearance of ever more complex and expensive diagnostic and therapeutic equipment added a new dimension. Many of these items are made of modern materials that cannot be resterilized by traditional means. This added to existing problems the need to invent and monitor new methods and apparatus for treating them to ensure that, at the very least, they are acceptably disinfected.

With these developments a new, increasingly sophisticated industry appeared in hospitals. The need for novel technical and managerial skills led to the development of special training programmes and the appearance of new paramedical specialists. Depending on circumstances, a variety of patterns (and nomenclatures) emerged. In some places a single central sterile supply department (CSSD) grew up to cater for the needs of wards, clinics, operating departments and sometimes for medical services in the community as well. In others separate services were provided, most often in the form of CSSDs plus theatre sterile supply units (TSSUs) that were often sited within operating departments. As more items appeared that had to be disinfected, CSSDs were sometimes renamed hospital sterilization and disinfection units (HSDUs).

The procedures carried out in a central sterilizing unit can be broken down into the reception of used equipment, which is then washed, dried, inspected and repacked together with whatever disposable items are required. Packs are then wrapped, sterilized and stored (often within a 'use by' period) until they are needed. A single layer of wrapping is sufficient if paper of the right quality is used and packs are handled with reasonable care. These procedures are modified as necessary to accommodate special items or equipment that can only withstand disinfection. In some places parts or the whole of the unit performing these functions are required to conform to clean-room conditions, with staff wearing lint-free clothing and working in HEPA-filtered air (p.218). This is another example of extreme and expensive measures being introduced intuitively, without proof of need.

Difficulties can arise when equipment returned to the central unit has been used on a patient who is judged to be 'infectious'. There may be concern that staff in the unit will be infected when they handle contaminated items during the early parts of the recycling process. The definition of what is 'infectious' has been a problem and what some people regard as dangerous may not be hazardous at all. A good example of an imaginary problem is MRSA, but concern about occupational infections with hepatitis B, perhaps C and to a lesser extent HIV cannot be dismissed so easily. Some centres have required that equipment used on an infected patient is only returned after either disinfection by clinical staff or actually being immersed in a disinfectant. The idea that this makes for greater safety ignores the fact that many patients with these infections are not identified as they pass through hospitals, so the precautions are not applied in every case (p.236). Although there is no record of a worker in a central unit being infected in this way, the concern is genuine and cannot be ignored.

Washer-disinfectors

Rather than using chemical disinfectants, which are messy, inefficient, ecologically unfriendly and expensive, it is better to pass all used instruments through a mechanical washer-disinfector shortly after receipt in the central unit, before they are handled at all. When this is done no special segregation is needed for 'infectious' items. The preliminary process cleans and simultaneously makes everything safe to handle, provided elementary care is taken. It also removes the need to clean and decontaminate used instruments before they reach the CSSD. The more expensive washer-sterilizers add nothing to the process. The unnecessary extra heat they provide acts to coagulate any blood or other proteinaceous materials, so sticks the dirt more firmly onto instruments.

A washer-disinfector is an extremely useful piece of machinery (Anon, 1983) which appears in hospitals in many guises. As the technology is the same as that found in a domestic dishwasher, it is well established. It operates on a principle similar to that used in laundries (p.246) in that a cool or cold rinse is followed by a hot wash, probably with a detergent, and the process ends with a hot rinse at about 80°C for between one and two minutes. The final temperature varies and so does the time. Figures of 65°C for ten minutes to 90°C for a few seconds have been quoted. The point is that provided the washing cycle is really efficient, the pasteurizing final rinse has little left to do except make the object being treated so hot that it conveniently dries itself on exposure to colder air. Machines employing this general principle are used in wards for cleaning and disinfecting bedpans and urinals, elsewhere for processing plastic or rubber anaesthetic equipment, in kitchens for cleaning and disinfecting eating utensils and in CSSDs as just indicated.

Of course, when purchasing equipment of this sort the claims of manufacturers should not be accepted at face value and machines should be checked both for their washing efficiency and the time and temperature achieved in the rinse cycle. Once installed, they need to be checked regularly for mechanical function and the temperature they reach. For this purpose cheap easily portable electronic thermometers are available. These instruments are attached by fine wires to small probes that can be inserted into the chambers of machines while they are operating. Some models are equipped with a variety of probes that can be used for different applications, including the measurement of airflows. Slightly more sophisticated models include a recorder so that a permanent record may be made over a period of time. Infection control teams should have access to such apparatus and its members should know how to use it.

KITCHENS

The preparation of food in hospitals is especially difficult and dangerous. The reasons for this are discussed and basic rules for the avoidance of food poisoning are described.

Introduction

Those responsible for hospital catering face unique problems. They must try to produce and distribute a palatable, safe diet to a great many people who are scattered through one or more large buildings, most of whom cannot assemble at central points and some of whom have particular and perhaps complex dietary requirements. The service must be available 24 hours a day, 365 days a year, is subject to budgetary restriction and may be provided from inadequate, outdated kitchens. Many of the clients are patients, who are by definition already sick. Sick people are more likely than the healthy to develop food-related illnesses and to die from them. The margin for error is much reduced. This scene is set in a medical environment where the results of even minor mistakes are much more likely to be noticed, investigated and reported. Because there is often a laboratory on the site it is more probable that any microbiological cause will be damningly identified. It is not surprising that outbreaks of food poisoning seem to be more common in hospitals than elsewhere.

Understandable it may be, but not excusable, Microbial food poisoning is avoidable and results from elementary mistakes in food hygiene. Most food poisoning in hospitals is of bacterial origin and the organisms responsible (often salmonellas or *Clostridium perfringens*) are nearly always present in small numbers in the food when it arrives at the hospital. The other main cause, *Staphylococcus aureus*, is usually added to the food in the kitchen, also in small numbers, by carriers of the bacterium (who comprise up to 50% of the population; p.46) or by someone suffering from sepsis due to it. Salmonella infections in hospitals are discussed further in Chapter 7, p.161.

The importance of temperature

A surprisingly large proportion of the raw food offered for sale internationally contains small numbers of food-poisoning microbes. More of it is contaminated in the kitchen, as described. In either case proper handling of the food can prevent this rather common contamination from becoming dangerous. The reason is that, as with all poisons, it is not so much the presence of the poison that matters but rather how much there is of it. In the case of bacterial food poisoning in more or less normal people it is necessary to have at least many thousands of the toxic bacteria in each gram of food before the poisonous threshold is reached. Bacteria can multiply in most kinds of food, but only at certain temperatures. For nearly all the bacteria that matter, multiplication is most rapid between about 20° and 40°C. Below freezing (food freezer temperature, nominally –18°C to –20°C) multiplication stops altogether. Between just above freezing and 8°C (food refrigerator temperatures) multiplication is so slow as to be effectively stopped in the short term (a few days). This is why *effective* refrigeration is such a vital part of food hygiene. An overloaded or inefficient refrigerator may be unable to achieve a safe temperature under any circumstances and anything hot that is put into it will take an unacceptably long time to chill. In either case bacterial multiplication will continue for much longer than it should.

Paradoxically, refrigeration encourages two foodborne pathogens. *Listeria monocytogenes* and *Yersinia enterocolitica* continue to multiply slowly in the refrigerator (particularly when the temperature rises above 4°C or 5°C), so that food contaminated with either of these bacteria may in time come to contain an infectious dose of them. At the same time effective chilling keeps the food 'wholesome' by preventing the growth of food-spoiling organisms. The epidemiology of the infections caused by these bacteria will continue to unfold as more and more mass-produced chilled food is distributed internationally. Because *L. monocytogenes* is an important pathogen for the immunocompromised it may emerge as a problem in hospitals. *Y. enterocolitica* has already caused problems because it can multiply in packs of blood refrigerated prior to transfusion.

As the temperature of food rises to between 40°C and 50°C bacterial multiplication ceases and above 50°C heat begins to become lethal, though of course less so for the spores of *C. perfringens*. Thorough cooking is very destructive of bacteria, but it must be remembered that large joints of meat or poultry take longer to heat up. This is particularly true if they have been frozen and are not properly defrosted before cooking begins. Carefully processed food that is eaten immediately after adequate heating is very safe from a microbiological point of view. It is unfortunate that thorough cooking makes some food unpalatable. Eating thus becomes a form of Russian roulette as some pathogenic microbes (and the enterotoxin of *Staph. aureus*) may escape destruction in the course of a brief exposure to heat.

Problems emerge if cooked food is stored prior to consumption. As the temperature of the food falls below about 40°C any surviving bacteria start to multiply again. If the temperature stays in the 'danger zone' (10°C–40°C) for long enough (an hour or two may be sufficient) the toxic threshold can be reached and those who consume the food will suffer from food poisoning. This can be avoided if heated food is kept hot (above 60°C) or if it is chilled rapidly to refrigerator temperature. This is a particular problem in hospitals because meals have to be distributed in relatively small amounts to many destinations. In these circumstances it is difficult to keep food at the desired temperature. The answer lies in timing, discipline and efficient refrigerators and food trolleys. Of course, food that has to be reheated (as happens in the cook-chill catering system) runs a double risk of being mishandled (Barrie, 1996).

Kitchen hygiene

The rules are simple and if they were obeyed food poisoning would be rare. They may be broken for a variety of reasons, including complaints about overcooked food, pressure of work, inadequate facilities or the employment of poorly trained staff with a rapid turnover. Food that is eaten raw or nearly so escapes the disinfecting effect of heat. Contamination of fruit and salads with gastrointestinal pathogens does not reach toxic levels if they have been properly handled and refrigerated. Eggs may be contaminated with small numbers of salmonellas when they are laid. If they are kept at room temperature and are then eaten raw or inadequately cooked, this can lead to food poisoning. When raw eggs are used as ingredients in foods that are not cooked prior to consumption (in mayonnaise, for example) even quite brief exposure to room temperature can convert a small dose of salmonellas into a toxic one. Because it may have grown in water contaminated by sewage, shellfish is always a gamble. Salads and the like should, of course, be kept cold and carefully washed before they are eaten.

Other points require attention. Separate refrigerators should be provided for fresh and cooked foods, to prevent the recontamination of a cooked item by contact with an uncooked one. All equipment, and in particular slicers, mixers, blenders and other catering machinery, together with work surfaces, must be kept scrupulously clean, using an approved catering detergent or disinfectant with plenty of hot water. Grubby cloths should be banned and the cutting, chopping and mixing components of catering machinery usually need to be taken apart to clean them properly.

The general hygiene of the kitchen and of the staff must be considered. A clean kitchen is the hallmark of a staff who are professionally competent and not under dangerous pressure. The fact that insects and other parasites are of low importance other than as indicators of low standards was discussed in Chapter 3, p.49. Staff suffering from diarrhoea or with septic skin lesions should not be allowed to work. Practice varies on the performance of routine laboratory examinations of the faeces of employees on first appointment or subsequently. Although this should always be done to clear an employee for return to work after an attack of diarrhoea, the answer otherwise varies with the incidence of gastrointestinal infections in the community from which the employees come. When this is low, routine examinations are not cost-effective. Examination may be limited to the taking of a medical history and the ordering of cultures only in cases where there has been a recent stomach upset or a history suggestive of enteric fever at any time. When the incidence of gastrointestinal infections (and infestations) is high, and in particular where typhoid or paratyphoid are common and if the facilities exist, faecal examinations ought to be done in every case.

One of the duties of the ICT is to visit and inspect hospital kitchens from time to time. If catering is done by a contractor, the ICC or ICT should be involved when tenders are invited to check that contracts are drawn up properly from the hygiene point of view (Barrie, 1996).

LAUNDRY

The process of laundering is described, with special attention paid to the perceived importance of the microbiological safety of the product.

Introduction

Towels, bedding, clothing, furnishings and the like make up the largest group of items used in healthcare facilities today that remain substantially untouched by the boom in disposables. The requirement is that, after use, they are treated by a process that renders them aesthetically acceptable (that is, clean and properly presented for reuse) and safe (particularly microbiologically so), but that does not unreasonably shorten their life. This is usually achieved by a process of washing in water followed by drying and pressing; that is, by laundering. Water is replaced by other solvents in the process of 'dry' cleaning.

The laundry cycle consists of one or more cold or cool rinses, followed by washing at a raised temperature, further rinsing, water extraction, drying, pressing, folding and packing. Such things as detergents, bleaches, conditioners, starches and so on are added at various points. The whole process may be done by hand or mechanized up to the level represented by a modern domestic washing machine or beyond to the level possible using the sophisticated machinery available to large commercial laundries.

Laundries

The initial cold or cool rinses wash off stains or other relatively soluble or loosely bound materials. This is particularly important if any of the dirt might be 'cooked' into the weave at a higher temperature. Natural egg-white, for example, is easily washed out of a piece of cloth in cold water, but it is denatured and becomes insoluble in hot water and is then very difficult to remove. The same applies to blood and other protein-containing body fluids and to a lesser extent to faeces. Washing at a raised temperature helps to extract less soluble materials and detergents, bleaches and enzymes or mixtures of these increase the efficiency of the process. The removal of dirt simultaneously removes microbes, which are subjected to dilution in the water, to the pasteurizing action of heat and to the disinfecting action of any chemicals that are added. Further significant decontamination may be achieved in the drying and pressing parts of the cycle, depending on the temperatures reached.

High-quality cotton and linen fabrics withstand washing at temperatures up to the boiling point of water, though many of the modern materials used to make clothing are damaged or destroyed at this temperature. This influences the choice of temperature used in the process. The amount of energy consumed should also be considered. It has been calculated that high-temperature laundering accounts for perhaps 15% of hospitals' total energy requirements. The cost of this, together with the non-replaceable nature of most current energy sources, imposes a need to use the lowest temperature that achieves the twin requirements of cleanliness and safety.

The hot part of the washing cycle may be performed at 71°C for a period of up to 30 minutes. Although vegetative bacteria and viruses (including HIV and almost certainly hepatitis B) are destroyed in less time, it is prolonged to ensure that all parts of the load in a large commercial washer reach the required temperature. The final temperature used varies, but for hot cycles it has usually fallen between 65°C and 80°C. Washing at lower temperatures would seem to risk less complete disinfection, but direct experiment has shown that when treated at temperatures as low as 40°C by a modern process in a modern machine products that are clean when they emerge are also substantially free of microbes (Blaser *et al.*, 1984; Tompkins, Johnson and Fittall, 1988). Clean laundry does not need to be sterile other than as drapes in operating rooms or for sterile packs and an autoclave is used to achieve this. For other purposes laundry that is aesthetically acceptable is likely to be microbiologically safe.

Clean and dirty laundry

Laundering is only a part of the continuous cycle through which fabrics pass until they are worn out. From the laundry items are returned to the linen store and issued to users in suitable containers that protect them from harm and keep them clean. What happens to them after use is more of a problem. Difficulties arise because some individuals perceive that soiled items returned to a laundry are a hazard to the people who work there. The danger has been greatly overstated. At the hospital end all that is required is to handle soiled items gently and pack them into impervious bags as quickly as possible. Gloves may be worn if the items are obviously contaminated with body fluids or have been used by a patient in isolation if required by hospital policy. Provided the bags are of reasonable quality double-bagging is not necessary. The question then arises, do soiled items need to be categorized by the risks they pose to laundry workers?

Risks might arise in laundries where soiled items are sorted by hand before they are put into washers. This is done to separate fabrics that call for different treatments and to remove items made of metal and other objects (used syringes, scissors, forceps, bowls, drugs, etc.) that are rather frequently found in bags of hospital laundry. Misplaced used syringes are a hazard to those who handle laundry at any stage and the metal items can cause expensive damage to the moving parts of washing machines. Although there are no reports of infections in operatives who perform this task (compare CSSDs, p.240), sorting items in this way carries a theoretical risk of occupational infection. In an attempt to minimize the problem hospital laundry was at one time categorized according to the level of risk thought to accompany it. Laundry may be 'soiled' (used but not objectionably dirty or dangerous), 'foul' (soiled with blood or faeces for example) or 'infected' (an impossibility as laundry is not alive, but meaning contaminated with microbes thought to be hazardous). It may be foul and infected simultaneously. Persons who do the sorting should wear good quality gloves, a waterproof apron, boots and perhaps eye protection. They should have been trained to be careful and be provided with proper changing and handwashing or showering facilities. If the individual can be trusted to follow instructions, there is almost no danger, even at a theoretical level.

Contaminated and potentially hazardous laundry

The duty of care imposes a constraint, however and it is difficult not to have a residual list of situations requiring greater precautions. These should not extend to infections such as those with MRSA, the hepatitis B virus or HIV, though emotion can overcome logic when laundry from cases of the latter begin to appear in a laundry for the first time. If smallpox still existed this would certainly require extra precautions, as would some rare conditions like Lassa, Marburg or Ebola fevers, anthrax and plague. The newly re-emerging problem of tuberculosis may also require special consideration. Laundry soiled with the faeces of patients with dysentery or cholera might also be treated differently. Infections that are specially hazardous to staff, including the staff of laundries, may be defined as 'quarantinable' and be subject to extra precautions. These are justified and may be applied more widely if the people who sort the linen at the laundry cannot be trusted to use the simple safety measures described in the last paragraph, which are sufficient to protect hospital staff from the same hazard (pp.188 and 195). In some hospitals linen that falls into the 'foul' category and is very wet is also dealt with differently, particularly if the bags used for ordinary laundry are not waterproof. Healthcare establishments must decide how far to extend these special precautions, remembering that they have costly consequences. Infection control organizations should be involved when these decisions are made.

In some places the special category of contaminated ('infected' or quarantinable) linen, however defined, is packed in bags made of plastic that dissolves in either cold or hot water or that are stitched together with soluble plastic thread. Laundry contained in these is put directly into washing machines without prior opening or sorting. It is then released into the water when the bag (or its stitching) dissolves. Depending on the type of plastic employed, this happens either in the cold or the hot part of the cycle.

Cold water-soluble bags or bags with cold water-soluble stitching have the problem that if items are wet when they are placed in them or if the bags are exposed to water (rain, for example) when they are full they may disintegrate, releasing their contents in inappropriate places. This is unacceptable. Hot water-soluble bags overcome this problem in part, but when using them it is sensible to wrap wet linen inside dry items. The temperature at which these bags dissolve seems to depend to some extent on the length of time for which they are exposed to water at any temperature. Their contents may still escape unexpectedly, particularly in warm and humid conditions.

Even when they work perfectly these bags suffer the disadvantages of high cost and the fact that their contents are not rinsed during the cold or cool part of the laundering cycle. There is a risk that any foul laundry put into them will be stained indelibly if, for example, blood is cooked into fabrics when bags release their contents directly into hot water. Provided the amount of laundry defined as hazardous is small, this risk may be accepted. To avoid the difficulty altogether it has been suggested that hot water-soluble bags closed with ties made of cold water-soluble material are used. To reduce the cost of the bags themselves it might seem attractive to use cheaper materials stitched together with hot water-soluble thread. Unfortunately the insoluble plastic of which the rest of the bag is made goes through the laundering process and can cause malfunctions if it gets caught in moving parts or it may soften and melt in the drying or pressing phases of the cycle. If these special bags are to be used at all their cost requires the development of good laundry policies, discipline in their application and a short list of unusual infections for which they are needed.

A sensible alternative to the use of soluble bags is to pack linen requiring special treatment in distinctive bags made of waterproof material. These may also be used for items that are unusually wet. If desired, the contents of these bags can be emptied into a washing machine without any preliminary sorting. An old solution to the problem of foul linen was to sluice it in the ward before it was bagged or to take it in specially identified bags to a central 'foul washer' somewhere in the hospital where it was rinsed in cold water prior to despatch to the laundry. In modern conditions neither of these practices is acceptable.

An important part of a laundry policy is to define and specify the types of laundry bag to be used. It is customary to have these made in certain easily distinguished colours, sometimes chosen on a regional or national basis. This is helpful when staff move between hospitals. Depending on the number of categories into which laundry is separated two, three or more types of bag are needed. A minimum is one for ordinary soiled linen and another for hazardous ('infected' or quarantinable) items, plus a third for fabrics to be washed at lower temperatures if this is an option and clothing is sorted at ward level. To avoid confusion the colours are chosen not to coincide with those used for bagging hospital waste (p.200).

DOMESTIC CLEANING

A dirty healthcare establishment suggests a low quality of care. It is the low quality of care that causes infections, not the dirt.

Introduction

Healthcare establishments should be clean and bright. A clean environment makes patients feel better and staff work more efficiently. Dirty hospitals (to be distinguished from old and shabby ones) can be dangerous places. This is not because dirt causes disease but because dirty hospitals, like dirty kitchens, are a sure sign of a poorly trained, sloppy staff and inadequate facilities. Infections and dirt are both products of low standards. A direct cause-and-effect relationship between dirt and disease exists mainly in the minds of those who confuse the real causes of infection with aesthetic considerations.

The indirect relationship noted between dirt and disease holds when the people who care for patients are also responsible for the cleanliness of their surroundings. This is no longer always so. A separation of roles has sometimes been imposed to increase commercial competitiveness or it has been the result of cost cutting in centrally funded services. It is now possible to have a squeaky-clean hospital with low-grade clinical staff, poor diagnostic services and too much IaI or, more often, a dirty hospital where dedicated and competent doctors and nurses work, backed up by good diagnostic departments, with a low incidence of IaI.

The dry environment contributes very little as a cause of IaI (p.51 and 52). Of course, the psychological importance of cleanliness allows no compromise, but routine environmental disinfection adds nothing useful. Expensive and environmentally unfriendly chemical disinfectants are almost never needed. In ordinary circumstances water and a detergent are all that is required to clean floors, walls and other surfaces throughout healthcare facilities, including operating rooms. From an infection control point of view the nature of the cleaning equipment does not matter so long as it does not raise too much dust. What does matter is that mops, squeegees and so on are washed clean after use, dried and stored dry. Some people feel more comfortable if mop-heads are autoclaved from time to time. There is no reason to think this is of any use. If mechanical floor scrubbers or, in particular, vacuum cleaners are used the air discharged from the machine should, on balance, be filtered. This is to avoid disseminating any of the more hardy pathogens that can be spread by the environmental airborne route (p.51 and 52). Note that the liquid in hand-sprayers used in the process of floor polishing by the 'spray-buff' method may support the growth of *Pseudomonas* spp. Hand-sprayers should not be 'topped up' with fresh polish, but carefully washed and dried between fillings.

Spills of body fluids

The hazards associated with spills of body fluids have been greatly overstated. They may be aesthetically objectionable, but they are rarely dangerous. Even if a spill contains large numbers of microbial pathogens the risk of infection to other patients or staff is low or very low, provided elementary precautions are taken. This is because microbes have to find a route by which to reach a portal of entry to cause an infection. Microbes contained in more or less viscous body fluids cannot escape into the air without the application of a great deal of force. Gentle handling using appropriate apparatus while wearing cheap non-sterile plastic gloves or even plastic bags on the hands makes the process safe. Unwarranted anxiety has led some to attempt to disinfect a spill before cleaning it away. Because the microbes that are most feared are viruses, chlorine is often recommended for the purpose. This has been combined with absorbent granules in a commercial product that simultaneously disinfects and soaks up the liquids in a spill. Otherwise, hypochlorite at the right concentration may be used (p.121). Disinfectants do not act instantaneously so if they are necessary at all (and there is serious doubt about this) logic decrees that the disinfectant must act for about ten minutes before anything else is done. In the meantime, the spill can be covered with, for example, absorbent paper towels. Final disposal is into the kind of plastic bag used for contaminated waste (p.204). With a liquid spill (of blood in particular) the preliminary disinfection may be dispensed with if the material is allowed to soak into paper towels. In operating rooms soiled drapes may be used for this purpose, if they are absorbent. If disinfectant has been applied to a spill logic again decrees that the area from which the spill has been removed is treated with fresh disinfectant before a final clean with detergent and water. If chlorine has been used in a confined space adequate ventilation must be ensured.

Many infection control organizations have concluded that the use of disinfectants for this purpose is an example of expensive overreaction and have abandoned it. It may be that the disinfection of spills can only be justified in laboratories and probably in the context of tuberculosis (see below).

Terminal disinfection

'Terminal disinfection' was part of the decontamination ritual of old fever hospitals. Unnecessary ritual still lingers in the procedures sometimes used when patients are discharged from isolation. Other than in cases of the more hazardous quarantinable diseases (as defined by local infection control organizations) nothing more than routine detergent-and-water cleaning is required (though again, see below for tuberculosis). Walls do not need to be cleaned unless they are dirty. Screens or curtains should not be laundered unless they are soiled. Bed linen that is not contaminated with faeces or urine needs no special treatment, provided the laundry is efficient (p.246). Books and other items do not need to be fumigated. Most pathogens survive poorly in a dry environment (p.50).

This apparently relaxed attitude is something of an oversimplification. The general impression that wiping a surface removes all particulate matter is supported by the commonplace observation that visible dust is readily swept away, particularly if a damp cloth or disposable wipe is used. In fact, this is not true of very small particles like bacteria, which adhere to surfaces much more firmly and which may in addition lurk in cracks and the microscopic crevices or scratches that mark even a smooth object. For this reason a significant number of the bacteria that were on a surface beforehand remain on it after it has been damp-dusted. If a cloth that has remained damp for an hour or two after earlier use is employed the action is likely to 'paint on' more bacteria than it removes. This is because in the interval bacteria, especially Gram-negative rods, have multiplied in the cloth. What happens next depends on how much water is left behind at the end of the process. If this is minimal, drying soon begins to kill the bacteria left behind. If a surface is left wet or if the damp-dusting is repeated frequently (as may be the case in busy kitchens) significant bacterial multiplication may take place in what has now become a wet environment (p.229). 'Clean and dry' is the watchword and when this is adhered to, little harm will follow.

Tuberculosis

Cases of open tuberculosis (patients not yet treated in whom the site of a tuberculous infection communicates with the outside world) are different. As noted on p.52, tubercle bacilli have a remarkable ability to withstand desiccation. In fact, careful washing with detergent and water will usually prevent their being resuspended in the air, which is the only place where they can do any harm. When cleaning is finished the dirty water should be disposed of into a proper sewage system and the wipe dealt with as clinical waste (p.200). With tuberculosis, however, the safety margins are narrow and there is a distant chance that bacteria left behind after wiping may survive for long enough so that eventually they are carried into the air. This becomes important when the number of tubercle bacilli is large (for example, where infected patients do not dispose of their sputum hygienically) and where there is no proper sewage system and in particular where open pulmonary tuberculosis is common in association with cases of AIDS. In these circumstances a phenolic disinfectant or a hypochlorite-detergent preparation at a concentration suitable for dirty situations (p.121) may be applied, before ordinary washing. The spread of multiresistant tubercle bacilli has made this a subject of growing importance and concern. A lack of precise information and the need to placate popular opinion may lead regulatory bodies to impose more extreme measures (p.196).

Staff safety

With the important exception of needle-stick injuries (pp.200 and 223), domestic employees in healthcare establishments are only at slightly more risk of infection than members of the general community. Healthcare workers may, without knowing it, have already met the patient they now see in hospital with a diagnosis of tuberculosis, hepatitis B or HIV infection. The previous meeting might have been in a store, on an escalator or in public transport. It is awareness, not the risk, that is new. Patients just admitted with illnesses not yet fully diagnosed are potentially just as dangerous. They may be undetected carriers of one of these infections and this explains why it is now recommended that all patients are approached on the basis that this may be so. With this in mind and if proper care is taken, for example by the use of simple protective clothing (pp.188 and 195), the risk is very small. Of course, when vaccines are available personnel should be vaccinated against the most serious hazards, in particular hepatitis B and perhaps tuberculosis as well (p.233).

Compared with employees in most other industries, HCWs work in a safe environment, but the fear of infection is a powerful instinct. Alarm can spread out of ignorance or may be deliberately fomented as a ploy when negotiating for improved pay or conditions of service. Education is the answer, provided the educator is properly informed and motivated. Of course, there is no excuse for lack of care by management and there is always room for improvement. It is as much a part of the duties of an ICT or an ICC to act in the interests of staff as to be concerned for patients.

REFERENCES

Anon (1983) Disinfection in washing machines. *Journal of Hospital Infection*, **4**, 101–2.

Barrie, D. (1996) The provision of food and catering services in hospital. *Journal of Hospital Infection*, **33**, 13–33.

Blaser, M. J., Smith, P. F., Cody, H. J. *et al.* (1984) Killing of fabric-associated bacteria in hospital laundry by low-temperature washing. *Journal of Infectious Diseases*, **149**, 48–57.

Heptonstall, J. and Mortimer, P. P. (1995) New virus, old story. *Lancet*, **345**, 599–600.

Monthly Report (1995) AIDS and HIV-1 infection worldwide. *Communicable Disease Report*, **5**, 97–8.

Tompkins, D. S., Johnson, P. and Fittall, B. R. (1988) Low-temperature washing of patients' clothing; effects of detergent with disinfectant and a tunnel drier on bacterial survival. *Journal of Hospital Infection*, **12**, 51–8.

Working Party (1995) *HIV and the Practice of Pathology*, Royal College of Pathologists, London.

What next in infection control? 12

INTRODUCTION

The prevention of infection in healthcare establishments has been a matter for concern for at least 150 years. This is not so much because attempts at control have failed but, notably, because they have succeeded. Many advances (the development of modern surgery, for example) were possible only when the infections that originally inhibited progress were controlled. The fact that infections are still a problem reflects the rapid expansion of medical science. No end is in sight as new diagnostic and therapeutic technologies bring new challenges to the control of infection.

This is not the whole story, for three reasons. First, most healthcare establishments are deficient in their practice of infection control because they do not apply all that is known about it. This may be due to lack of education, lack of information or lack of motivation. Second, some of the things that are done are wasteful, unnecessary or both. Infection control practices need to be reviewed from time to time to see if they are effective and cost effective. Third, the approach to infection control is less scientific than it should be. More logic and less woolly thought will make it easier to identify and cut out useless rituals and avoid new mistakes. Attention to each of these areas will influence the future development of infection control.

DEFICIENCIES IN PRACTICE, EDUCATION AND SCIENCE

By far the most common problem is a lack of resources. In wards where there are no handbasins or if soap and towels are not provided, calls for hand hygiene fall on deaf ears. This is also attributable to ignorance. The control of infection will receive a fairer share of available resources when its case is supported with facts that reflect local conditions. It is difficult to impress managers with data derived by extrapolation from surveys made by others thousands of miles away, many years ago (Box 5.1).

It is not uncommon to find people in the healthcare field who are so ignorant about iatrogenic infections that they may even deny that they exist. If the importance of these infections is underestimated, little or nothing will be done to prevent them. Even when this is not true, busy people may fail to keep abreast of developments in the field of infection control. This is more likely to happen if the fact that the information exists is unknown or if it is inaccessible. Even when the information does dribble through it may be misinterpreted by those who lack the basic knowledge about infection required for its comprehension. These kinds of difficulties are widespread and may coexist.

The scientific practice of infection control has developed in a spasmodic manner, based to a large extent on individual initiatives. Too often the science has been diluted by exhortation based on intuitive reasoning and degraded by erroneous ideas from the past. There exists a considerable body of misinformation. The proliferation of independent vocational groups within the healthcare industry has made this more of a problem than it used to be. Several of these groups have sought and gained professional autonomy and have exercised their right to educate new recruits to their ranks. Misplaced professional pride has led to unhealthy inbreeding among the educators appointed from within some of these groups. The growth of knowledge in medicine is such that nobody can grasp all of it. Training in appropriate scientific disciplines is needed by those who are to interpret or evaluate developments within specialized fields. To understand the control of infection requires some competence in the related fields of epidemiology, infectious diseases and microbiology. In addition to the basic information necessary, those who teach infection control require access to current literature in the field and the time to digest it. Vocational instructors who include in their teaching what for them is the unfamiliar subject of infection control may unwittingly use sources that are unreliable and out of date. If they do not call on specialists in infection control to help them they will almost certainly perpetuate error and mislead their pupils.

The same difficulty affects members of the various groups who move into senior management, where they must make major policy decisions. Where these impinge on infection control mistakes are made by those whose knowledge of the subject depends on the imperfect recollection of possibly erroneous teaching a long time ago. It is only necessary to walk round new hospitals to see the errors that are made. The need to control infection is often a secondary consideration in major policy decisions. Managers and administrators who are perfectly able to make the primary decisions need to be aware of their limitations in these secondary areas. If they do not turn to the right quarter for advice they are likely to make surprisingly expensive errors.

It has been suggested elsewhere that in a perfect world each doctor and nurse would know enough about infection control to make specialists in the field redundant. This state of perfection is unlikely to be achieved for a long time. In the meantime the deficiencies noted will be dealt with most efficiently by education, both primary and in-service (p.95). This should involve all professional groups, not forgetting the educators themselves. Well-trained infection control doctors (ICDs) and infection control nurses (ICNs) are primary resources in this respect. The skills required by these people are described in Chapter 5, p.87. Formal training courses for ICNs exist in many places. To avoid the unhealthy inbreeding just described these should always involve multidisciplinary teachers. This is particularly important because specialist knowledge to the level required is not taught in ordinary nurse-training programmes. Most of the training of ICDs is done by apprenticeship to an established practitioner. In some centres short formal courses are also available. Again, these are important to avoid inbreeding.

COST-EFFECTIVENESS

A recurring theme in this book has been the need to question what is done to control infection to see if the practices are soundly based or if some of them need to be modified or abandoned as ineffective or illogical, or both. The object of infection control is to reduce the morbidity and mortality associated with healthcare, without wasting money. Practices vary enormously between hospitals and this alone must indicate that some of them are unnecessary. Every ICT and ICC needs to audit the rates of infection in the medical establishments under their control and examine the effectiveness of the control measures they apply (p.66). They are likely to find areas where they ought to do better (for example, in the use and care of urinary catheters) and locate pointless and expensive rituals (the overuse of disinfectants or unnecessary elaboration in dressing wounds, for example). These monitoring activities will keep infection controllers busy for a long time and it will not be long before failure to include such formal audit among their activities is counted as a major deficiency.

THE SCIENTIFIC APPROACH

Infection control resembles a stool with three legs. The legs represent the disciplines of epidemiology, infectious diseases and microbiology. Without all three legs the stool will fall over and it will not provide a firm seat if any of them is deficient.

Semmelweis showed that problems can be solved by the application of epidemiological techniques to the study of infectious diseases (p.6). This approach takes time, years in Semmelweis' case. When the unit of measurement is death, time is expensive. It is interesting to speculate what would have happened if Semmelweis had been able to call on microbiology to help him. This ought to have made his task easier and saved many lives. On the other hand he might have been diverted from his classic epidemiological approach when he found (as he would have done) that the environment in his wards was hopping with streptococci. This is what happened to the bacteriologists in the 1930s whose failure to be rigorous in their epidemiological thinking resulted in the establishment of illogical rituals, some of which still persist, and a considerable waste of money (p.5).

On the other hand the story of legionnaire's disease (legionellosis) is an example of how epidemiology can be hampered by the absence of microbiology. It was not possible to make the elaborately documented epidemiological facts of the famous Philadelphia outbreak mean anything until *Legionella pneumophila* had been isolated and its habits elucidated. The isolation took about six months to achieve. At a much more mundane level, the need to distinguish between clinically indistinguishable infections due to different bacteria or different types of the same microbe (p.34) underlines the importance of the modern relationship between epidemiology and microbiology. Because a study must start with accurate definitions and diagnoses, the third leg (a knowledge of infectious diseases) is also critical to a successful outcome.

If three people are required to provide these three forms of expertise then there are three points of interpersonal contact at which communication can fail or friction develop. Other things being equal, and for economy, the smaller the number of people needed to bring together the three skills, the better. However, anyone who wishes to practise infection control effectively must recognize the importance of balancing this triple relationship. If less than three people are involved, care must be taken to see that one or more of the legs of the stool is not underdeveloped. A medical approach to disease is helpful. Microbiology without epidemiology will get nowhere and epidemiology on its own can produce masses of data of little or no value. Only so much information should be collected as can be put to good use, either in the education of people so they behave better or to act as the basis for properly designed studies of the causes or prevention of infection. If too much time is spent collecting data there may be no time left to use them. To collect infection control data only for use in litigation is a sad comment on the modern practice of medicine.

It is clear that there is enormous potential for development in infection control. This will be achieved if practitioners base their activities firmly in science, though without forgetting their humanity. Nothing is more caring than the control of infection, but this should depend on logical persuasion rather than on flights of imagination and rhetorical exhortation. Windy rhetoric frequently conceals an intellectual void!

Appendix
Definitions of infections

An essential part of any epidemiological study is a careful definition of the phenomenon to be measured (p.71). The reason for this is that if a collection of data includes phenomena not related to the one being surveyed the analysis will be confused and may defeat the object of the exercise. In a study of yellow fever in South America the researchers inadvertently included cases of leptospirosis, as both can cause jaundice. As a result the researchers were confused, the wrong conclusion was drawn and a lot of hard work and money were wasted. Their definition of yellow fever was insufficiently rigorous. When counting apples, pears must be excluded!

In the case of iatrogenic infections clear, unambiguous definitions are even more important because comparison will be made between data that have been collected by different people at different times in different places. Nurses or other paramedical staff are often required to collect raw IaI data and so have to make the sometimes difficult distinction between colonizations and infections. On the definitions themselves even experts sometimes disagree. It is not easy to please everybody!

A set of definitions may occupy a single sheet or cover many pages. A fuller example comes from CDC Atlanta (Garner, J.S. *et al.* (1988) CDC definitions for nosocomial infections, *American Journal of Infection Control*, **16**, 128–40). These, designed for incidence surveys, are complex and offer complicated alternatives. They require careful study. Another set of definitions for incidence surveys is provided in Glenister H. M. *et al.* (1992, An 11-month incidence study of infection in the wards of a district general hospital. *Journal of Hospital Infection*, **21**, 261–73). At the other end of the spectrum are the brief definitions used in a prevalence survey of hospital-acquired infections in England and Wales in 1980 (Meers, P.D. *et al.* (1981) Report on the national survey of infection in hospitals, 1980. *Journal of Hospital Infection*, **2**, (supplement), 48–51). These were prepared for use by separate teams in each of the 43 hospitals surveyed. The teams consisted of each hospital's control of infections officer (CIO) (usually a medically qualified microbiologist) and an infection control nurse (ICN), together with a senior nurse from each of the wards visited. One member of the team had been trained for the task beforehand and a team, sometimes accompanied by an external scrutineer, saw each of the 18 186 patients involved. A somewhat modified form of the definitions that were used in this survey are reproduced below. A more recent set for another prevalence survey may be found in the Report of a Steering Group (1993, National prevalence survey of hospital acquired infections: definitions. *Journal of Hospital Infection*, **24**, 69–76).

As with most choices concerned with infection control, local decisions about definitions are best made by the ICC of the hospital that intends to use them or, for a more general survey, by an appropriate group appointed for the purpose. It is necessary to take account of previous definitions and to modify these to suit their requirements. The form in which they are written must reflect the type of survey intended and the background training of those who are to use them.

THE DEFINITIONS: GENERALIZATIONS

- In an **infection** a patient is reacting clinically or subclinically to the presence of a pathogenic microbe. In the absence of a reaction the condition is a **colonization**.
- A **iatrogenic infection** is an infection found in a patient who is or who has been under medical care and that is a consequence of that care. (This definition may be expanded to include infections in healthcare workers that result from their employment and may have to be modified when applied to iatrogenic infections in the community; see Chapter 9.)
- A **self-infection** is caused by a microbe that was part of the colonizing or normal flora of the patient *before* an infection due to it was initiated. When the microbe was part of the patient's normal flora prior to the institution of medical care the infection is an **endogenous** self-infection. When a colonization with a 'hospital' pathogen has developed after admission to care and this pathogen subsequently causes an infection, it is an **autogenous** self-infection.
- A **cross-infection** is an infection due to a microbe that came from another patient, a member of staff or a visitor, in a healthcare establishment.
- An **environmental infection** is an infection due to a microbe that originated in the inanimate environment of the patient.

THE DEFINITIONS: SPECIFIC INFECTIONS

- A **urinary tract infection** is recorded when a patient is under treatment for a microbiologically or clinically diagnosed infection of the urinary tract or in whom a firm diagnosis has been made but no treatment is given.
- A **respiratory tract infection** is recorded as upper or lower. In an **upper infection** the patient has acute coryzal symptoms or a significant inflamed mouth, throat, sinus or middle ear, not due to allergy. In a **lower infection**, there is new or increased purulent sputum production with chest signs and/ or X-ray changes not attributable to a non-infectious cause (see also p.148).
- A **wound infection**: a wound is a break in an epithelial surface (skin or mucous membrane) made by some act, such as an accident, burn or a surgical incision. (Infected ulcers and pressure sores may be recorded either as wound or skin infections.) A wound is infected if there is a purulent discharge in or exuding from it. Infections may be classified as **minor** if the integrity of a surgical wound is not threatened or **major** if the wound seems

likely to break down, or has done so. Surgical wounds may be categorized as follows (see also p.135): **clean**, a surgical incision into tissue that is not inflamed and the gastrointestinal, respiratory or genital tracts are not entered; **clean-contaminated**, an otherwise clean wound which entered one of the above systems, so that microbial contamination might occur, but where no significant spillage was observed or was likely; **contaminated**, where one of the above systems has been opened and microbial contamination was observed or was probable or where an area of inflammation was encountered; **dirty**, where pus was encountered, a perforated viscus found or in operations on contaminated traumatic wounds.

- A **skin infection**: skin and subcutaneous tissues that display the classic signs of inflammation, not apparently due to a chemical or other non-infectious cause (see wound infections).
- **Bacteraemia**: positive blood cultures, excluding those due to contamination, without symptoms.
- **Septicaemia**: as for bacteraemia, with symptoms or, in the absence of a culture, when a firm clinical diagnosis of septicaemia has been made.
- Infections of **ears, eyes, genital tract**: the presence of inflammation or of significant new purulent discharges.
- **CNS infections**: a positive culture from or microscopy of cerebrospinal fluid. Symptoms and signs of an abscess or of encephalitis.
- **Gastrointestinal infection** may be generalized or localized. **Generalized**: appropriate symptoms, plus a report of the recognition of a gastro-intestinal pathogen, if part of a recognized outbreak, a case with characteristic symptoms. **Localized**: a clinical diagnosis of appendicitis, diverticulitis, cholecystitis, anorectal sepsis, dental abscess, etc.
- Infections of **bones and joints**: adequate evidence of osteomyelitis or septic arthritis.
- **Other infections**: any other obvious infection, including classic ones such as measles, hepatitis, etc.
- In every case due weight is given to any **microbiological evidence** that is available, but only to the extent that a culture or other laboratory result is supported by information from other sources and it is consistent with the circumstances.

Index

Entries in **bold** refer to Figures, in *italic* to Tables, and in ***bold italic*** to Boxes